D1196289

Practical Heat Treating

by
Howard E. Boyer

American Society for Metals
Metals Park, OH 44073

First printing, April 1984
Second printing, March 1985
Third printing, March 1987
Fourth printing, June 1991

Library of Congress Cataloging in Publication Data

Boyer, Howard E.
  Practical heat treating.

  Includes index.
  1. Metals—Heat treatment.   I. Title.
TN672.B68   1984   671.3'6   83-25881
ISBN 0-87170-178-2

SAN 204-7586

Printed in the United States of America

Cover photo: Courtesy of Babcock & Wilcox Insulating Products Division, Augusta, Georgia. Shows a 27 foot long by 13-1/2 foot wide carbottom furnace at Georgia Iron Works.

Preface

Practical Heat Treating presents heat treating technology in clear, concise, and nontheoretical language. It is directed to design engineers, manufacturing engineers, shop personnel, students, and others requiring an understanding of why heat treatment is specified and how the various heat treating processes are employed to obtain desired engineering properties.

Mr. Boyer provides this fundamental information by first explaining briefly the principles of the heat treatment of steel and the concepts of hardness and hardenability. Consideration is given next to furnaces and related equipment. The major portion of the book, however, is devoted to a discussion of the commonly used heat treatments for carbon and alloy steels, tool steels, stainless steels, and cast irons. Sample treatments are given in detail for many of the commercially important and commonly specified grades. Chapters on case hardening procedures, flame and induction heating and the heat treating of nonferrous alloys complete the book.

Mr. Boyer, past Editor of the *Metals Handbook*, called upon 50 years of practical experience when writing this book. His editorial approach centers on the basic information a person must know and understand to perform common engineering functions. The reader looking for a fundamental explanation of heat treating technology will find *Practical Heat Treating* an invaluable reference.

Contents

Chapter 1

What Is Heat Treatment? Classification and Commercial Importance

The generally accepted definition for heat treating metals and metal alloys is "heating and cooling a solid metal or alloy in a way so as to obtain specific conditions and/or properties." Heating for the sole purpose of hot working (as in forging operations) is excluded from this definition. Likewise, the types of heat treatment that are sometimes used for products such as glass or plastics are also excluded from coverage by this definition.

Commercial Importance

It would be difficult to imagine what life would be like if the properties of metals could not be altered in a variety of ways through the use of heat treatment. Without the benefits derived from heat treating, the auto industry, airplane/aerospace industries, and countless everyday hardware items would be nonexistent. Indeed, a sharp-edge steel razor blade would not be available!

Almost all metals and alloys respond to some form of heat treatment, in the broadest sense of the definition. The response of various metals and alloys, however, is by no means equal. Almost any pure metal or alloy can be softened (annealed) by means of a suitable heating and cooling cycle; however, the number of alloys that can be strengthened or hardened by heat treatment is far more restricted.

Practically all steels respond to one or more type of heat treatment. This is the major reason why steels account for over 80% of total metal production. The underlying principles of the heat treatment of steel are discussed in Chapter 2.

Many nonferrous alloys—namely aluminum, copper, nickel, magnesium, and titanium alloys—can be strengthened to various degrees by specially designed heat treatments, but not to the same degree and not by the same techniques as steel. The heat treatment of nonferrous metals is covered in Chapter 12.

Classification of Heat Treating Processes

In some instances, heat treatment procedures are clear cut in terms of technique and application, whereas in other instances, descriptions or simple explanations are insufficient because the same technique frequently may be used to obtain different objectives. For example, stress relieving and tempering are often accomplished with the same equipment and by use of identical time and temperature cycles. The objectives, however, are different for the two processes.

The following descriptions of the principal heat treating processes are generally arranged according to their interrelationships.

Normalizing consists of heating a ferrous alloy to a suitable temperature (usually 50 to 100 °F, or 28 to 56 °C) above its specific upper transformation temperature (see Chapter 2 and the glossary of terms at the end of this chapter). This is followed by cooling in still air to at least some temperature well below its transformation temperature range. For low-carbon steels, the resulting structure and properties are the same as those achieved by full annealing; for most ferrous alloys, normalizing and annealing are not synonymous.

Normalizing usually is used as a conditioning treatment, notably for refining the grains of steels that have been subjected to high temperatures for forging or other hot working operations. The normalizing process usually is succeeded by another heat treating operation such as austenitizing for hardening, annealing, or tempering.

Annealing is a generic term denoting a heat treatment that consists of heating to and holding at a suitable temperature followed by cooling at a suitable rate. It is used primarily to soften metallic materials, but also to simultaneously produce desired changes in other properties or in microstructure. The purpose of such changes may be, but is not confined to, improvement of machinability, facilitation of cold work (known as process, or in-process annealing), improvement of mechanical or electrical properties, or to increase dimensional stability. When applied solely to relieve stresses, it commonly is called stress-relief annealing, synonymous with stress relieving.

When the term "annealing" is applied to ferrous alloys without qualification, full annealing is implied. This is achieved by heating above the alloy's transformation temperature, then applying a cooling cycle which provides maximum softness. This cycle may vary widely, depending on composition and characteristics of the specific alloy.

In nonferrous alloys, annealing cycles are designed to (1) remove part or all of the effects of cold working (recrystallization may or may not be involved), (2) cause substantially complete coalescence of precipitates from solid solution in relatively coarse form, or (3) both, depending on composition and material condition. Specific process names in commercial use are final annealing, full annealing, intermediate annealing, partial annealing, recrystallization annealing, and stress-relief annealing. Some of these are "in-shop" terms that do not have precise definitions.

Austenitizing is defined as the process of forming austenite by heating a ferrous alloy above the transformation range. When used without qualification, the term implies complete austenitizing. Under no circumstances is austenitizing a complete process, but rather it is the initial step prior to normalizing, full annealing of ferrous alloys, or quench-hardening of ferrous alloys. Austenite is the phase of a ferrous alloy which generally exists only at elevated temperature. For a more complete explanation of austenite and its formation, see Chapter 2.

Quenching is the rapid cooling of a steel or alloy from the austenitizing temperature by immersing the workpiece in a liquid or gaseous medium. Quenching media commonly used include water, 5% brine, 5% caustic in an aqueous solution, oil, polymer solutions, or gas (usually air or nitrogen).

Selection of a quenching medium depends largely on the hardenability of the material (see Chapter 3) and the mass of the material being treated (principally section thickness).

The cooling capabilities of the above-listed quenching media vary greatly. In selecting a quenching medium, it is best to avoid a solution that has more cooling power than is needed to achieve the results, thus minimizing the possibility of cracking and warpage of the parts being treated. Modifications of the term quenching include direct quenching, fog quenching, hot quenching, interrupted quenching, selective quenching, spray quenching, and time quenching. These terms are defined in the glossary at the end of this chapter.

Tempering. In heat treating of ferrous alloys, tempering consists of reheating the austenitized and quench-hardened steel or iron to some preselected temperature that is below the lower transformation temperature (generally below 1300 °F or 705 °C). Tempering offers a means of obtaining various combinations of mechanical properties. Tempering temperatures used for hardened steels are often no higher than 300 °F (150 °C). The term "tempering" should not be confused with either process annealing or stress relieving. Even though time and temperature cycles for the three processes may be the same, the conditions of the materials being processed and the objectives may be different.

Stress Relieving. Like tempering, stress relieving is always done by heating to some temperature below the lower transformation temperature for steels and irons. For nonferrous metals, the temperature may vary from slightly above room temperature to several hundred degrees, depending on the alloy and the amount of stress relief that is desired.

The primary purpose of stress relieving is to relieve stresses that have been imparted to the workpiece from such processes as forming, rolling, machining, or welding. The usual procedure is to heat workpieces to the pre-established temperature long enough to reduce the residual stresses (this is a time-and temperature-dependent operation) to an acceptable level; this is followed by cooling at a relatively slow rate to avoid creation of new stresses.

Carburizing consists of absorption and diffusion of carbon into solid ferrous alloys by heating to some temperature above the upper transformation temperature of the specific alloy. Temperatures used for carburizing are generally in the range of 1650 to 1900 °F (900 to 1040 °C). Heating is done in a carbonaceous environment (liquid, solid, or gas). This produces a carbon gradient extending inward from the surface, enabling the surface layers to be hardened to a high degree either by quenching from the carburizing temperature or by cooling to room temperature followed by reaustenitizing and quenching. Carburizing is discussed in greater detail in Chapter 10.

Carbonitriding is a case hardening process in which a ferrous material (most often a low-carbon grade of steel) is heated above the transformation temperature in a gaseous atmosphere of such composition as to cause simultaneous absorption of carbon and nitrogen by the surface and, by diffusion, create a concentration gradient. The process is completed by cooling at a rate that produces the desired properties in the workpiece.

Carbonitriding is most widely used for producing thin, hard, wear-resistant cases on numerous hardware items. It is well adapted to mass production, and is sometimes referred to as "gas cyaniding" because the results are similar to those produced by a molten salt bath that contains a substantial amount of cyanide (see Cyaniding below). For more details on carbonitriding, see Chapter 10.

Cyaniding is also a case hardening process wherein a ferrous metal (usually, although not necessarily, a low-carbon steel) is heated above its transformation temperature; that is, the work material is "austenitized" in a molten salt that contains cyanide. The result is simultaneous absorption of carbon and nitrogen at the surface, and by diffusion, a concentration gradient is created. Quenching completes the formation of a hard, wear-resistant case with a relatively soft interior.

The hardness and structure resulting from the cyaniding process are at least similar if not identical with those obtained by carbonitriding. Carbonitriding has, however, replaced cyaniding to a large extent because (1) disposal of the cyanide salts is difficult, and (2) it is difficult to remove the residual salts from cyanide-hardened workpieces, especially those of intricate design.

Nitriding. The introduction of nitrogen into the surface layers of certain ferrous alloys by holding at a suitable temperature below the lower transformation temperature, (Ac$_1$) in contact with a nitrogenous environment is known as nitriding. The introduction of nitrogen into the surface layers of certain range of 975 to 1050 °F (525 to 565 °C), and the nascent nitrogen may be generated by cracking of anhydrous ammonia (NH$_3$) or from molten salts that contain cyanide. Quenching is not required to create a hard, wear-resistant and heat-resistant case (see Chapter 10 for more details on the nitriding process).

Nitrocarburizing is a relatively new term and is used to describe an entire family of processes by which both nitrogen and carbon are absorbed into the surface layers of a wide variety of carbon and alloy steels. The sources for carbon and nitrogen may be either molten salt or gas, and temperatures are generally below the lower transformation temperature of the alloy; that is, below Ac$_1$. Nitrocarburizing not only provides a wear-resistant surface, but also increases fatigue strength. Retention of these properties depends largely on the avoidance of finishing operations after the heat treatment because nitrocarburized cases are extremely thin. There are a number of proprietary processes that comprise the family of nitrocarburizing processes (see Chapter 10 for further details on and applications of nitrocarburizing).

Glossary of Heat Treating Terms

The most commonly used heat treating terms are defined and/or described below. For a complete listing of metallurgical terms, see Ref 1.

Ac$_{cm}$, Ac$_1$, Ac$_3$, Ac$_4$. Defined under *transformation temperature*.

A$_{cm}$, A$_1$, A$_3$, A$_4$. Same as Ae$_{cm}$, Ae$_1$, Ae$_3$, and Ae$_4$.

acicular ferrite. A highly substructured nonequiaxed ferrite that forms upon continuous cooling by a mixed diffusion and shear mode of transformation that begins at a temperature slightly higher than the temperature transformation range for upper bainite. It is distinguished from bainite in that it has a limited amount of carbon available; thus, there is only a small amount of carbide present.

Ae$_{cm}$, Ae$_1$, Ae$_3$, Ae$_4$. Defined under *transformation temperature*.

age hardening. Hardening by aging, usually after rapid cooling or cold working. See *aging*.

aging. A change in the properties of certain metals and alloys that occurs at ambient or moderately elevated temperatures after hot working or a heat treatment (quench aging in ferrous alloys, natural or artificial aging in ferrous and nonferrous alloys) or after a cold working operation (strain aging). The change in properties is often, but not always, due to a phase change (precipitation), but never involves a change in chemical composition of the metal or alloy. See also *age hardening, natural aging, precipitation hardening, precipitation heat treatment.*

allotropy. A near synonym for *polymorphism.* Allotropy is generally restricted to describing polymorphic behavior in elements, terminal phases, and alloys whose behavior closely parallels that of the predominant constituent element.

alloy steel. Steel containing specified quantities of alloying elements (other than carbon and the commonly accepted amounts of manganese, copper, silicon, sulfur, and phosphorus) within the limits recognized for constructional alloy steels, added to effect changes in mechanical or physical properties.

austempering. A heat treatment for ferrous alloys in which a part is quenched from the austenitizing temperature at a rate fast enough to avoid formation of ferrite or pearlite and then held at a temperature just above M_s until transformation to bainite is complete.

austenite. A solid solution of one or more elements in face-centered cubic iron. Unless otherwise designated (such as nickel austenite), the solute is generally assumed to be carbon.

austenitic steel. An alloy steel whose structure is normally austenitic at room temperature.

bainite. A metastable aggregate of ferrite and cementite resulting from the transformation of austenite at temperatures below the pearlite range but above M_s. Its appearance is feathery if formed in the upper part of the bainite transformation range; acicular.

banded structure. A segregated structure consisting of alternating nearly parallel bands of different composition, typically aligned in the direction of primary hot working.

bright annealing. Annealing in a protective medium to prevent discoloration of the bright surface.

Brinell hardness test. A test for determining the hardness of a material by forcing a hard steel or carbide ball of specified diameter into it under a specified load. The result is expressed as the Brinell hardness number, which is the value obtained by dividing the applied load in kilograms by the surface area of the resulting impression in square millimetres.

carbon equivalent. (1) For cast iron, an empirical relationship of the total carbon, silicon, and phosphorus contents expressed by the formula:

$$CE = TC + 1/3(Si + P)$$

(2) For rating of weldability:

$$CE = C + \frac{Mn}{6} + \frac{Cr + Mo + V}{5} + \frac{Ni + Cu}{15}$$

carbonization. Conversion of an organic substance into elemental carbon. (Should not be confused with carburization.)

carbon potential. A measure of the ability of an environment containing active carbon to alter or maintain, under prescribed conditions, the carbon level of the steel. Note: In any particular environment, the carbon level attained will depend on such factors as temperature, time and steel composition.

carbon restoration. Replacing the carbon lost in the surface layer from previous processing by carburizing this layer to substantially the original carbon level. Sometimes called recarburizing.

carbon steel. Steel having no specified minimum quantity for any alloying element (other than the commonly accepted amounts of manganese, silicon and copper) and that contains only an incidental amount of any element other than carbon, silicon, manganese, copper, sulfur, and phosphorus.

case hardening. A generic term covering several processes applicable to steel that change the chemical composition of the surface by absorption of carbon or nitrogen, or both, and by diffusion create a concentration gradient. See *carburizing, cyaniding, carbonitriding, nitriding, and nitrocarburizing.*

cast iron. A generic term for a large family of cast ferrous alloys in which the carbon content exceeds the solubility of carbon in austenite at the eutectic temperature. Most cast irons contain at least 2% C plus silicon and sulfur, and may or may not contain other alloying elements. For the various forms *gray cast iron, white cast iron, malleable cast iron*, and *ductile cast iron,* the word "cast" is often left out, resulting in "gray iron," "white iron," "malleable iron," and "ductile iron," respectively.

cementite. A compound of iron and carbon, known chemically as iron carbide and having the approximate chemical formula Fe_3C. It is characterized by an orthorhombic crystal structure. When it occurs as a phase in steel, the chemical composition will be altered by the presence of manganese and other carbide-forming elements.

cold treatment. Exposing to suitable subzero temperatures for the purpose of obtaining desired conditions or properties such as dimensional or microstructural stability. When the treatment involves the transformation of retained austenite, it is usually followed by tempering.

combined carbon. The part of the total carbon in steel or cast iron that is present as other than *free carbon.*

constitution diagram. A graphical representation of the temperature and composition limits of phase fields in an alloy system as they actually exist under the specific conditions of heating or cooling (synonymous with phase diagram). A constitution diagram may be an equilibrium diagram, an approximation to an equilibrium diagram, or a representation of metastable conditions or phases. Compare with *equilibrium diagram.*

critical temperature. (1) Synonymous with *critical point* if the pressure is constant. (2) The temperature above which the vapor phase cannot be condensed to liquid by an increase in pressure.

critical temperature ranges. Synonymous with *transformation ranges*, which is the preferred term.

decarburization. Loss of carbon from the surface layer of a carbon-containing alloy due to reaction with one or more chemical substances in a medium that contacts the surface.

direct quenching. (1) Quenching carburized parts directly from the carburizing operation. (2) Also used for quenching pearlitic malleable parts directly from the malleablizing operation.

double aging. Employment of two different aging treatments to control the type of precipitate formed from a supersaturated matrix in order to obtain the desired properties. The first aging treatment, sometimes referred to as intermediate or stabilizing, is usually carried out at higher temperature than the second.

ductile cast iron. A *cast iron* that has been treated while molten with an element such as magnesium or cerium to induce the formation of free graphite as nodules or spherulites, which imparts a measurable degree of ductility to the cast metal. Also known as nodular cast iron, spherulitic graphite cast iron and SG iron.

end-quench hardenability test. A laboratory procedure for determining the hardenability of a steel or other ferrous alloy; widely referred to as the Jominy test. Hardenability is determined by heating a standard specimen above the upper critical temperature, placing the hot specimen in a fixture so that a stream of cold water impinges on one end, and, after cooling to room temperature is completed, measuring the hardness near the surface of the specimen at regularly spaced intervals along its length. The data are normally plotted as hardness versus distance from the quenched end.

equilibrium. A dynamic condition of physical, chemical, mechanical or atomic balance, where the condition appears to be one of rest rather than change.

equilibrium diagram. A graphical representation of the temperature, pressure and composition limits of phase fields in an alloy system as they exist under conditions of complete equilibrium. In metal systems, pressure is usually considered constant.

eutectic carbide. Carbide formed during freezing as one of the mutually insoluble phases participating in the eutectic reaction of ferrous alloys.

eutectic melting. Melting of localized microscopic areas whose composition corresponds to that of the eutectic in the system.

fatigue strength. The maximum stress that can be sustained for a specified number of cycles without failure, the stress being completely reversed within each cycle unless otherwise stated.

ferrimagnetic material. A material that macroscopically has properties similar to those of a *ferromagnetic material* but that microscopically also resembles an antiferromagnetic material in that some of the elementary magnetic moments are aligned antiparallel. If the moments are of different magnitudes, the material may still have a large resultant magnetization.

ferrite. (1) A solid solution of one or more elements in body-centered cubic iron. Unless otherwise designated (for instance, as chromium ferrite), the solute is generally assumed to be carbon. On some equilibrium diagrams there are two ferrite regions separated by an austenite area. The lower area is alpha ferrite; the upper, delta ferrite. If there is no designation, alpha ferrite is assumed. (2) In the field of magnetics, substances having the general formula:

$$M^{++}O \cdot M_2^{+++}O_3,$$

the trivalent metal often being iron.

ferritizing anneal. A treatment given as-cast gray or ductile (nodular) iron to produce an essentially ferritic matrix. For the term to be meaningful, the final microstructure desired or the time-temperature cycle used must be specified.

flame annealing. Annealing in which the heat is applied directly by a flame.

flame hardening. A process for hardening the surfaces of hardenable ferrous alloys in which an intense flame is used to heat the surface layers above the upper transformation temperature, whereupon the workpiece is immediately quenched.

fog quenching. Quenching in a fine vapor or mist.

free carbon. The part of the total carbon in steel or cast iron that is present in elemental form as graphite or temper carbon. Contrast with *combined carbon.*

free ferrite. Ferrite that is formed directly from the decomposition of hypoeutectoid austenite during cooling, without the simultaneous formation of cementite. Also proeutectoid ferrite.

full hard. A *temper* of nonferrous alloys and some ferrous alloys corresponding approximately to a cold worked state beyond which the material can no longer be formed by bending. In specifications, a full hard temper is commonly defined in terms of minimum hardness or minimum tensile strength (or, alternatively, a range of hardness or strength) corresponding to a specific percentage of cold reduction following a full anneal. For aluminum, a full hard temper is equivalent to a reduction of 75% from *dead soft*; for austenitic stainless steels, a reduction of about 50 to 55%.

gamma structure. A Hume-Rothery designation for structurally analogous phases or electron compounds that have ratios of 21 valence electrons to 13 atoms; generally, a large complex cubic structure. Not to be confused with gamma phase on a constitution diagram.

gamma iron. The face-centered cubic form of pure iron, stable from 910 to 1400 °C (1670 to 2550 °F).

graphitic steel. Alloy steel made so that part of the carbon is present as graphite.

graphitization. Formation of graphite in iron or steel. Where graphite is formed during solidification, the phenomenon is called primary graphitization; where formed later by heat treatment, secondary graphitization.

graphitizing. Annealing a ferrous alloy in such a way that some or all of the carbon is precipitated as graphite.

gray cast iron. A *cast iron* that gives a gray fracture due to the presence of flake graphite. Often called gray iron.

half hard. A *temper* of nonferrous alloys and some ferrous alloys characterized by tensile strength about midway between that of *dead soft* and *full hard* tempers.

hardenability. The relative ability of a ferrous alloy to form martensite when quenched from a temperature above the upper critical temperature. Hardenability is commonly measured as the distance below a quenched surface where the metal exhibits a specific hardness (50 HRC, for example) or a specific percentage of martensite in the microstructure.

hardness. Resistance of metal to plastic deformation, usually by indentation. However, the term may also refer to stiffness or temper, or to resistance to scratching, abrasion or cutting. Indentation hardness may be measured by various hardness tests, such as *Brinell, Rockwell,* and *Vickers.*

hypereutectic alloy. In an alloy system exhibiting a *eutectic*, any alloy whose composition has an excess of alloying element compared to the eutectic composition, and whose equilibrium microstructure contains some eutectic structure.

hypereutectoid alloy. In an alloy system exhibiting a *eutectoid*, any alloy whose composition has an excess of alloying element compared to the eutectoid composition, and whose equilibrium microstructure contains some eutectoid structure.

hypoeutectic alloy. In an alloy system exhibiting a *eutectic*, any alloy whose composition has an excess of base metal compared to the eutectic composition, and whose equilibrium microstructure contains some eutectic structure.

hypoeutectoid alloy. In an alloy system exhibiting a *eutectoid*, any alloy whose composition has an excess of base metal compared to the eutectoid composition, and whose equilibrium microstructure contains some eutectoid structure.

induction hardening. A surface-hardening process in which only the surface layer of a suitable ferrous workpiece is heated by electromagnetic induction to above the upper critical temperature and immediately quenched.

induction heating. Heating by combined electrical resistance and hysteresis losses induced by subjecting a metal to the varying magnetic field surrounding a coil carrying alternating current.

interrupted quenching. A quenching procedure in which the workpiece is removed from the first quench at a temperature substantially higher than that of the quenchant and is then subjected to a second quenching system having a different cooling rate.

interstitial solid solution. A solid solution in which the solute atoms occupy positions that do not correspond to lattice points of the solvent. Contrast with *substitutional solid solution.*

isothermal annealing. Austenitizing a ferrous alloy and then cooling to and holding at a temperature at which austenite transforms to a relatively soft ferrite carbide aggregate.

isothermal transformation. A change in phase that takes place at a constant temperature. The time required for transformation to be completed, and in some instances the time delay before transformation begins, depends on the amount of supercooling below (or superheating above) the equilibrium temperature for the same transformation.

Knoop hardness. Microhardness determined from the resistance of metal to indentation by a pyramidal diamond indenter, having edge angles of 172° 30′ and 130°, making a rhombohedral impression with one long and one short diagonal.

maraging. A precipitation-hardening treatment applied to a special group of iron-based alloys to precipitate one or more intermetallic compounds in a matrix of essentially carbon-free martensite. Note: the first developed series of maraging steels contained, in addition to iron, more than 10% Ni and one or more supplemental hardening elements. In this series, aging is done at 900 °F (480 °C).

malleable cast iron. A cast iron made by a prolonged anneal of *whi.e cast iron* in which decarburization or graphitization, or both, take place to eliminate some or all of the cementite. The graphite is in the form of temper carbon. If decarburization is the predominant reaction, the product will have a light fracture, hence, "whiteheart malleable;" otherwise, the fracture will be dark, hence, "blackheart malleable." Ferritic malleable has a predominently ferritic matrix; pearlitic malleable may contain pearlite, spheroidite or tempered martensite depending on heat treatment and desired hardness.

malleablizing. Annealing *white cast iron* in such a way that some or all of the combined carbon is transformed to graphite or, in some instances, part of the carbon is removed completely.

martempering. (1) a hardening procedure in which an austenitized ferrous workpiece is quenched into an appropriate medium whose temperature is maintained substantially at the M_s of the workpiece, held in the medium until its temperature is uniform throughout — but not long enough to permit bainite to form —and then cooled in air. The treatment is frequently followed by tempering. (2) When the process is applied to carburized material, the controlling M_s temperature is that of the case. This variation of the process is frequently called marquenching.

martensite. (1) In an alloy, a metastable transitional structure intermediate between two allotropic modifications whose abilities to dissolve a given solute differ considerably, the high-temperature phase having the greater solubility. The amount of the high-temperature phase transformed to martensite depends to a large extent upon the temperature attained in cooling, there being a rather distinct beginning temperature. (2) A metastable phase of steel, formed by a transformation of austenite below the M_s (or Ar″) temperature. It is an interstitial supersaturated solid solution of carbon in iron having a body-centered tetragonal lattice. Its microstructure is characterized by an acicular, or needlelike, pattern.

martensite range. The temperature interval between M_s and M_f.

martensitic transformation. A reaction that takes place in some metals on cooling, with the formation of an acicular structure called *martensite*.

metastable. Refers to a state of pseudoequilibrium that has a higher free energy than the true equilibrium state.

M_f temperature. For any alloy system, the temperature at which martensite formation on cooling is essentially finished. See *transformation temperature* for the definition applicable to ferrous alloys.

microhardness. The hardness of a material as determined by forcing an indenter such as a Vickers or Knoop indenter into the surface of a material under very light load; usually, the indentations are so small that they must be measured with a microscope. Capable of determining hardnesses of different microconstituents within a structure, or of measuring steep hardness gradients such as those encountered in case hardening.

M_s temperature. For any alloy system, the temperature at which martensite starts to form on cooling. See *transformation temperature* for the definition applicable to ferrous alloys.

natural aging. Spontaneous aging of a supersaturated solid solution at room temperature. See *aging*, and compare with *artificial aging* (aging above room temperature).

overheating. Heating a metal or alloy to such a high temperature that its properties are impaired. When the original properties cannot be restored by further heat treating, by mechanical working or by a combination of working and heat treating, the overheating is known as *burning*.

partial annealing. An imprecise term used to denote a treatment given cold worked material to reduce the strength to a controlled level or to effect stress relief. To be meaningful, the type of material, the degree of cold work, and the time-temperature schedule must be stated.

pearlite. A metastable lamellar aggregate of ferrite and cementite resulting from the transformation of austenite at temperatures above the bainite range.

phase. A physically homogeneous and distinct portion of a material system.

physical metallurgy. The science and technology dealing with the properties of metals and alloys, and of the effects of composition, processing and environment on those properties.

precipitation hardening. Hardening caused by the precipitation of a constituent from a supersaturated solid solution. See also *age hardening* and *aging*.

precipitation heat treatment. *Artificial aging* (aging above room temperature) in which a constituent precipitates from a supersaturated solid solution.

preheating. Heating before some further thermal or mechanical treatment. For tool steel, heating to an intermediate temperature immediately before final austenitizing. For some nonferrous alloys, heating to a high temperature for a long time, in order to homogenize the structure before working. In welding and related processes, heating to an intermediate temperature for a short time immediately before welding, brazing, soldering, cutting, or thermal spraying.

process annealing. An imprecise term denoting various treatments used to improve workability. For the term to be meaningful, the condition of the material and the time-temperature cycle used must be stated.

pusher furnace. A type of continuous furnace in which parts to be heated are periodically charged into the furnace in containers, which are pushed along the hearth against a line of previously charged containers thus advancing the containers toward the discharge end of the furnace, where they are removed.

quench annealing. Annealing an austenitic ferrous alloy by *solution heat treatment* followed by rapid quenching.

quench cracking. Fracture of a metal during quenching from elevated temperature. Most frequently observed in hardened carbon steel, alloy steel or tool steel parts of high hardness and low toughness. Cracks often emanate from fillets, holes, corners or other stress raisers and result from high stresses due to the volume changes accompanying transformation to martensite.

quench hardening. (1) Hardening suitable alpha-beta alloys (most often certain copper or titanium alloys) by solution treating and quenching to develop a martensite-like structure. (2) In ferrous alloys, hardening by austenitizing and then cooling at a rate such that a substantial amount of austenite transforms to martensite.

recrystallization. (1) The formation of a new, strain-free grain structure from that existing in cold worked metal, usually accomplished by heating. (2) The change from one crystal structure to another, as occurs on heating or cooling through a critical temperature.

recrystallization annealing. Annealing cold worked metal to produce a new grain structure without phase change.

recrystallization temperature. The approximate minimum temperature at which complete recrystallization of a cold worked metal occurs within a specified time.

Rockwell hardness test. An indentation hardness test based on the depth of penetration of a specified penetrator into the specimen under certain arbitrarily fixed conditions.

rotary furnace. A circular furnace constructed so that the hearth and workpieces rotate around the axis of the furnace during heating.

Scleroscope test. A hardness test where the loss in kinetic energy of a falling metal "tup", absorbed by indentation upon impact of the tup on the metal being tested, is indicated by the height of rebound.

selective quenching. Quenching only certain portions of an object.

shell hardening. A surface-hardening process in which a suitable steel workpiece, when heated through and quench hardened, develops a martensitic layer or shell that closely follows the contour of the piece and surrounds a core of essentially pearlitic transformation product. This result is accomplished by a proper balance among section size, steel hardenability, and severity of quench.

slack quenching. The incomplete hardening of steel due to quenching from the austenitizing temperature at a rate slower than the critical cooling rate for the particular steel, resulting in the formation of one or more transformation products in addition to martensite.

snap temper. A precautionary interim stress-relieving treatment applied to high-hardenability steels immediately after quenching to prevent cracking because of delay in tempering them at the prescribed higher temperature.

solution heat treatment. Heating an alloy to a suitable temperature, holding at that temperature long enough to cause one or more constituents to enter into solid solution, and then cooling rapidly enough to hold these constituents in solution.

spheroidizing. Heating and cooling to produce a spheroidal or globular form of carbide in steel. Spheroidizing methods frequently used are (1) prolonged holding at a temperature just below Ae_1; (2) heating and cooling alternately between temperatures that are just above and just below Ae_1; (3) heating to a temperature above Ae_1 or Ae_3 and then cooling very slowly in the furnace or holding at a temperature just below Ae_1.

steel. An iron-based alloy, malleable in some temperature ranges as initially cast, containing manganese, usually carbon, and often other alloying elements. In carbon steel and low-alloy steel, the maximum carbon is about 2.0%; in high-alloy steel, about 2.5%. The dividing line between low-alloy and high-alloy steels is generally regarded as being at about 5% metallic alloying elements.

Steel is to be differentiated from two general classes of irons: the cast irons, on the high-carbon side, and the relatively pure irons such as ingot iron, carbonyl iron, and electrolytic iron, on the low-carbon side. In some steels containing extremely low carbon, the manganese content is the principal differentiating factor, steel usually containing at least 0.25%; ingot iron, considerably less.

thermocouple. A device for measuring temperatures, consisting of lengths of two dissimilar metals or alloys that are electrically joined at one end and connected to a voltage-measuring instrument at the other end. When one junction is hotter than the other, a thermal electromotive force is produced that is roughly proportional to the difference in temperature between the hot and cold junctions.

time quenching. Interrupted quenching in which the time in the quenching medium is controlled.

total carbon. The sum of the free and combined carbon (including carbon in solution) in a ferrous alloy.

transformation temperature. The temperature at which a change in phase occurs. The term is sometimes used to denote the limiting temperature of a transformation range. The following symbols are used for iron and steels:

Ac_{cm}. In hypereutectoid steel, the temperature at which the solution of cementite in austenite is completed during heating.
Ac_1. The temperature at which austenite begins to form during heating.
Ac_3. The temperature at which transformation of ferrite to austenite is completed during heating.
Ac_4. The temperature at which austenite transforms to delta ferrite during heating.
Ae_{cm}, Ae_1, Ae_3, Ae_4. The temperatures of phase changes at equilibrium.
Ar_{cm}. In hypereutectoid steel, the temperature at which precipitation of cementite starts during cooling.
Ar_1. The temperature at which transformation of austenite to ferrite or to ferrite plus cementite is completed during cooling.

Ar_3. The temperature at which austenite begins to transform to ferrite during cooling.
Ar_4. The temperature at which delta ferrite transforms to austenite during cooling.
Ar'. The temperature at which transformation of austenite to pearlite starts during cooling.
M_f. The temperature at which transformation of austenite to martensite finishes during cooling.
M_s (or Ar''). The temperature at which transformation of austenite to martensite starts during cooling.

Note: All these changes except the formation of martensite occur at lower temperatures during cooling than during heating, and depend on the rate of change of temperature.

Vickers hardness test. An indentation hardness test employing a 136° diamond pyramid indenter (Vickers) and variable loads enabling the use of one hardness scale for all ranges of hardness from very soft lead to tungsten carbide.

Chapter 2

Fundamentals Involved in the Heat Treatment of Steel

Before consideration can be given to the heat treatment of steel or other iron-base alloys, it is helpful to explain what steel is. The common dictionary definition is "a hard, tough metal composed of iron, alloyed with various small percentages of carbon and often variously with other metals such as nickel, chromium, manganese, etc." Although this definition is not untrue, it is hardly adequate.

In the glossary of this book, the principal portion of the definition for steel is "an iron-base alloy, malleable in some temperature range as initially cast, containing manganese, usually carbon, and often other alloying elements. In carbon steel and low-alloy steel, the maximum carbon is about 2.0%; in high-alloy steel, about 2.5%. The dividing line between low-alloy and high-alloy steels is generally regarded as being at about 5% metallic alloying elements." (Ref 1)

Again, the definition is essentially correct; however when the reader completes this chapter, it will be evident that a comprehensive definition of this "wonder metal" is practically impossible.

Fundamentally, all steels are mixtures, or more properly, alloys of iron and carbon. However, even the so-called plain carbon steels have small, but specified, amounts of manganese and silicon plus small and generally unavoidable amounts of phosphorus and sulfur. The carbon content of plain carbon steels may be as high as 2.0%, but such an alloy is rarely found. Carbon content of commercial steels usually ranges from 0.05 to about 1.0%.

The alloying mechanism for iron and carbon is different from the more common and numerous other alloy systems in that the alloying of iron and carbon occurs as a two-step process. In the initial step, iron combines with 6.67% C, forming iron carbide which is called cementite. Thus at room temperature, conventional steels consist of a mixture of cementite and ferrite (essentially iron). Each of these is known as a phase (defined as a physically homogeneous and distinct portion of a material system). When a steel is heated above 1340 °F (725 °C), cementite dissolves in the matrix, and a new phase is formed which is called austenite. Note that phases of steel should not be confused with structures. There are only three phases involved in any steel—ferrite, carbide (cementite), and austenite, whereas there are several structures or mixtures of structures.

Classification of Steels

It is impossible to determine the precise number of steel compositions and other variations that presently exist, although the total number certainly exceeds 1000; thus, any rigid classification is impossible. However, steels are arbitrarily divided into five groups. which has proved generally satisfactory to the metalworking community.

These five classes are carbon steels, alloy steels, stainless steels, tool steels, and special-purpose steels. The first four of these groups are well defined by standards set forth by the American Iron and Steel Institute (AISI), and each general class is subdivided into numerous groups, with each grade identified. The fifth group is comprised of several hundred different compositions; most of them are proprietary. Many of these special steels are similar to specific steels in the first four groups, but vary sufficiently to be marketed as separate compositions. For example, AISI lists nearly 75 stainless steels in 4 different general subdivisions. In addition to these steels (generally referred to as "standard grades"), there are at least 100 other compositions that are nonstandard. Each was developed to serve a specific application.

For each of the other general classifications, there are many special-purpose steels that closely resemble one or more of the standard grades. Coverage in this book is limited to steels of the first four classes—carbon, alloy, stainless, and tool steels—that are listed by AISI. Space does not permit coverage of heat treatment practices of the nonstandard grades.

Why Steel Is a "Wonder Metal"

It would be unjust to state that any one metal is more important than another without defining parameters of consideration. For example, without aluminum and titanium alloys, airplanes and space vehicles would be nonexistent.

Steel, however, is by far the most widely used alloy and for a very good reason. Among laymen, the reason for steel's dominance is usually considered to be the abundance of iron ore (iron is the principal ingredient in all steels) and/or the ease by which it can be refined from ore. Neither of these is necessarily correct; iron is by no means the most abundant element, and it is not the easiest metal to produce from ore. Copper, for example, exists as nearly pure metal in certain parts of the world.

Steel is often called a "wonder metal" because of its tremendous flexibility in metal working and heat treating to produce a wide variety of mechanical, physical, and chemical properties.

Metallurgical Phenomena

The broad possibilities provided by the use of steel are attributed mainly to two all-important metallurgical phenomena: (1) iron is an allotropic element, that is it can exist in more than one crystalline form; and (2) the carbon atom is only 1/30th the size of the iron atom. These phenomena are thus the underlying principles which permit the achievements that are possible through heat treatment.

In entering the following discussion of constitution, however, it must be emphasized that a minimum of technical description is unavoidable. This portion of the subject is inherently technical. To avoid that would result in the discussion becoming uninformative and generally useless. The purpose of this chapter is, therefore, to reduce the prominent technical features toward their broadest generalizations and to present those generalizations and underlying principles in a manner that should instruct the reader interested in the metallurgical principles of steel. This is done at the risk of some oversimplification.

Constitution of Iron

It should first be made clear to the reader that any mention of molten metal is purely academic; this book deals exclusively with the heat treating range that is well below the melting temperature. The objective of this section is to begin with a generalized discussion of the constitution of commercially pure iron, subsequently leading to discussion of the iron-carbon alloy system which is the basis for all steels and their heat treatment.

All pure metals, as well as alloys, have individual constitutional or phase diagrams. As a rule, percentages of two principal elements are shown on the horizontal axis of a figure, while temperature variation is shown on the vertical axis. However, the constitutional diagram of a pure metal is a simple vertical line. The constitutional diagram for commercially pure iron is presented in Fig. 1. This specific diagram is a straight line as far as any changes are concerned, although time is indicated on the horizontal. As pure iron, in this case, cools, it changes from one phase to another at constant temperature. No attempt is made, however, to quantify time, but merely to indicate as a matter of interest that as temperature increases, reaction time decreases which is true in almost any solid solution reaction.

Pure iron solidifies from the liquid at 2800 °F (1538 °C) (top of Fig. 1). A crystalline structure, known as ferrite, or delta iron, is formed (point a, Fig. 1). This structure, in terms of atom arrangement, is known as a body-centered cubic lattice (bcc), shown in Fig. 2(a). This lattice has nine atoms—one at each corner and one in the center.

As cooling proceeds further and point b (Fig. 1) is reached (2541 °F or 1395 °C), the atoms rearrange into a 14-atom lattice as shown in Fig. 2(b).

The lattice now has an atom at each corner and one at the center of each face. This is known as a face-centered cubic lattice (fcc), and this structure is called gamma iron.

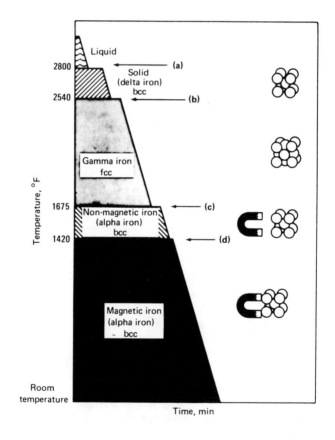

Fig. 1 Changes in pure iron as it cools from the molten state to room temperature. Source: MEI Course 1, Lesson 11, American Society for Metals, p 5, 1977.

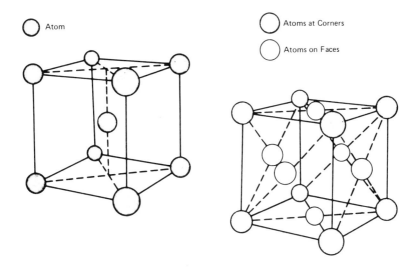

Fig. 2 Arrangement of atoms in the two crystalline structures of pure iron.

As cooling further proceeds to 1673 °F (910 °C) (point c, Fig. 1), the structure reverts to the nine-atom lattice or alpha iron. The change at point d on Fig. 1 (1418 °F or 770 °C) merely denotes a change from nonmagnetic to magnetic iron and does not represent a phase change. The entire field below 1673 °F (910 °C) is comprised of alpha ferrite, which continues on down to room temperature and below. The ferrite forming above the temperature range of austenite is often referred to as delta ferrite; that forming below A_3 as alpha ferrite, though both are structurally similar. In this Greek-letter sequence, austenite is gamma iron, and the interchangeability of these terms should not confuse the fact that only two structurally distinct forms of iron exist.

Figures 1 and 2 (Ref 2) thus illustrate the allotropy of iron. In the following sections of this chapter, the mechanism of allotropy as the all-important phenomenon relating to the heat treatment of iron-carbon alloys is discussed.

Alloying Mechanisms

Metal alloys are usually formed by mixing together two or more metals in their molten state. The two most common methods of alloying are (1) by atom exchange and (2) by the interstitial mechanism. The mechanism by which two metals alloy is greatly influenced by the relative atom size. The exchange mechanism simply involves trading of atoms from one lattice system to another. An example of alloying by exchange is the copper-nickel system wherein atoms are exchanged back and forth.

Interstitial alloying requires that there be a large variation in atom sizes between the elements involved. Because the small carbon atom is 1/30th the size of the iron atom, interstitial alloying is easily facilitated. Under certain conditions, the tiny carbon atoms enter the lattice (the interstices) of the iron crystal (Fig. 2). A description of this basic mechanism follows.

Effect of Carbon on the Constitution of Iron

As an elemental metal, pure iron has only limited engineering usefulness despite its allotropy. Carbon is the main alloying addition that capitalizes on the allotropic phenomenon and lifts iron from mediocrity into the position of the world's unique structural material, broadly known as steel. Even in the highly alloyed stainless steels, it is the quite minor constituent carbon which virtually controls the engineering properties. Furthermore, due to the manufacturing processes, carbon in effective quantities persists in all irons and steels unless special methods are used to minimize it.

Carbon is almost insoluble in iron, which is in the alpha or ferritic phase (1673 °F or 910 °C). However, it is quite soluble in gamma iron. Carbon actually dissolves; that is, the individual atoms of carbon lose themselves in the interstices among the iron atoms. Certain interstices within the fcc structure (austenite) are considerably more accommodating to carbon than are those of ferrite, the other allotrope. This preference exists not only on the mechanical basis of size of opening, however, for it is also a fundamental matter involving electron bonding and the balance of those attractive and repulsive forces which underlie the allotrope phenomenon.

The effects of carbon on certain characteristics of pure iron are shown in Fig. 3 (Ref 3). Figure 3(a) is a simplified version of Fig. 1; that is, a straight line constitutional diagram of commercially pure iron. In Fig. 3(b), the diagram is expanded horizontally to depict the initial effects of carbon on the principal thermal points of pure iron. Thus, each vertical dashed line, like the solid line in Fig. 3(a), is a constitutional diagram, but now for iron containing that particular percentage of carbon. Note that carbon lowers the freezing point of iron and that it broadens the temperature range of austenite by raising the temperature A_4 at which (delta) ferrite changes to austenite and by lowering the temperature A_3 at which the austenite reverts to (alpha) ferrite. Hence, carbon is said to be an austenitizing element. The spread of arrows at A_3 covers a two-phase region, which signifies that austenite is retained fully down to the temperatures of the heavy arrow, and only in part down through the zone of the lesser arrows.

In a practical approach, however, it should be emphasized that Fig. 1, as well as Fig. 3, represent changes that occur during very slow cooling, as would be possible during laboratory controlled experiments, rather than under

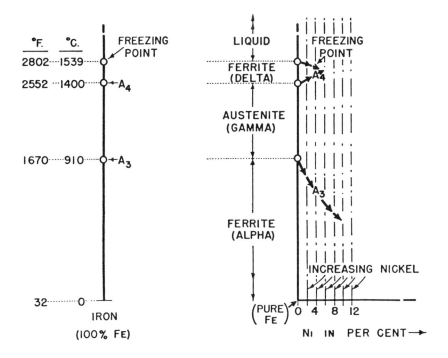

Fig. 3 Effects of carbon on the characteristics of commercially pure iron. Source: Zapffe, C.A., *Stainless Steels*, American Society for Metals, p 88, 1949.

conditions in commercial practice. Furthermore, in slow heating of iron, the above transformations take place in a reverse manner. Transformations occuring at such slow rates of cooling and heating are known as equilibrium transformations, due to the fact that sufficient time is allowed for transformations to occur at the temperatures indicated in Fig. 1.

Therefore, the process by which iron changes from one atomic arrangement to another when heated through 1673 °F (910 °C) is called a transformation. Transformations of this type occur not only in pure iron but also in many of its alloys; each alloy composition transforms at its own characteristic temperature. It is this transformation that makes possible the variety of properties that can be achieved to a high degree of reproducibility through use of carefully selected heat treatments.

Iron-Cementite Phase Diagram

When carbon atoms are present, two changes occur (see Fig. 3). First, transformation temperatures are lowered, and second, transformation takes place over a range of temperatures rather than at a single temperature. These data are shown in the well-known iron-cementite phase diagram (Fig. 4).

However, a word of explanation is offered to clarify the distinction between phases and phase diagrams.

A phase is a portion of an alloy, physically, chemically, or crystallographically homogeneous throughout, which is separated from the rest of the alloy by distinct bounding surfaces. Phases that occur in iron-carbon alloys are molten alloy, austenite (gamma phase), ferrite (alpha phase), cementite, and graphite. These phases are also called constituents. Not all constituents (such as pearlite or bainite) are phases—these are microstructures.

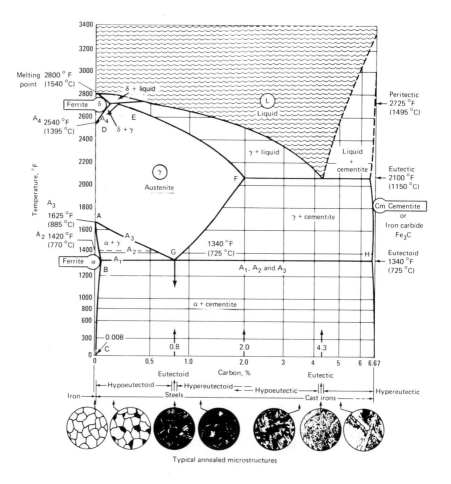

Fig. 4 Iron-cementite phase diagram. Source: MEI Course 1, Lesson 11, American Society for Metals, p 8, 1977.

A phase diagram is a graphical representation of the equilibrium temperature and composition limits of phase fields and phase reactions in an alloy system. In the iron-cementite system, temperature is plotted vertically, and composition is plotted horizontally. The iron-cementite diagram (Fig. 4), deals only with the constitution of the iron-iron carbide system, i.e., what phases are present at each temperature and the composition limits of each phase. Any point on the diagram, therefore, represents a definite composition and temperature, each value being found by projecting to the proper reference axis.

Although this diagram extends from a temperature of 3500 °F (1930 °C) down to room temperature, note that part of the diagram lies below 1900 °F (1040 °C). Steel heat treating practice rarely involves the use of temperatures above 1900 °F (1040 °C). In metal systems, pressure is usually considered as constant.

Frequent reference is made to the iron-cementite diagram (Fig. 4) in this chapter and throughout this book. Consequently, understanding of this concept and diagram is essential to further discussion.

The iron-cementite diagram is frequently referred to incorrectly as the iron-carbon equilibrium diagram. Iron-"carbon" is incorrect because the phase at the extreme right is cementite, rather than carbon or graphite; the term equilibrium is not entirely appropriate because the cementite phase in the iron-graphite system is not really stable. In other words, given sufficient time (less is required at higher temperatures), iron carbide (cementite) decomposes to iron and graphite, i.e., the steel graphitizes. This is a perfectly natural reaction, and only the iron-graphite diagram (see Chapter 9) is properly referred to as a true equilibrium diagram.

Solubility of Carbon in Iron

In Fig. 4, the area denoted as austenite is actually an area within which iron can retain much dissolved carbon. In fact, most heat treating operations (notably annealing, normalizing, and heating for hardening) begin with heating the alloy into the austenitic range to dissolve the carbide in the iron. At no time during such heating operations are the iron, carbon, or austenite in the molten state. A solid solution of carbon in iron can be visualized as a pyramidal stack of basketballs with golf balls between the spaces in the pile. In this analogy, the basketballs would be the iron atoms, while the golf balls interspersed between them would be the smaller carbon atoms.

Austenite is the term applied to the solid solution of carbon in gamma iron, and, like other constituents in the diagram, austenite has a certain definite solubility for carbon, which depends on the temperature (shaded area in Fig.

4 bounded by AGFED). As indicated by the austenite area in Fig. 4, the carbon content of austenite can range from 0 to 2%. Under normal conditions, austenite cannot exist at room temperature in plain carbon steels; it can exist only at elevated temperatures bounded by the lines AGFED in Fig. 4. Although austenite does not ordinarily exist at room temperature in carbon steels, the rate at which steels are cooled from the austenitic range has a profound influence on the room temperature microstructure and properties of carbon steels. Thus, the phase known as austenite is fcc iron, capable of containing up to 2% dissolved carbon.

The solubility limit for carbon in the bcc structure of iron-carbon alloys is shown by the line ABC in Fig. 4. This area of the diagram is labeled alpha (α), and the phase is called ferrite. The maximum solubility of carbon in alpha iron (ferrite) is 0.025% and occurs at 1340 °F (725 °C). At room temperature, ferrite can dissolve only 0.008% C, as shown in Fig. 4. This is the narrow area at the extreme left of Fig. 4 below approximately 1670 °F (910 °C). For all practical purposes, this area has no effect on heat treatment and shall not be discussed further. Further discussion of Fig. 4 is necessary, although as previously stated, the area of interest for heat treatment extends vertically to only about 1900 °F (1040 °C) and horizontally to a carbon content of 2%. The large area extending vertically from zero to the line BGH (1340 °F or 725 °C) and horizontally to 2% C is denoted as a two-phase area—α + Cm, or alpha (ferrite) plus cementite (carbide). The line BGH is known as the lower transformation temperature (A_1). The line AGH is the upper transformation temperature (A_3). The triangular area ABG is also a two-phase area, but the phases are alpha and gamma, or ferrite plus austenite. As carbon content increases, the A_3 temperature decreases until the eutectoid is reached—1340 °F (725 °C) and 0.80% C (point G). This is considered a saturation point; it indicates the amount of carbon that can be dissolved at 1340 °F (725 °C). A_1 and A_3 intersect and remain as one line to point H as indicated. The area above 1340 °F (725 °C) and to the right of the austenite region is another two-phase field—gamma plus cementite (austenite plus carbide).

Now as an example, when a 0.40% carbon steel is heated to 1340 °F (725 °C), its crystalline structure begins to transform to austenite; transformation is not complete however until a temperature of approximately 1500 °F (815 °C) is reached. In contrast, as shown in Fig. 4, a steel containing 0.80% C transforms completely to austenite when heated to 1340 °F (725 °C). Now assume that a steel containing 1.0% C is heated to 1340 °F (725 °C) or just above. At this temperature, austenite is formed, but because only 0.80% C can be completely dissolved in the austenite, 0.20% C remains as cementite, unless the temperature is increased. However, if the temperature of a 1.0% carbon steel is increased above about 1450 °F (790 °C), the line GF is intersected, and all of the carbon is thus dissolved. Increasing temperature gradually increases the amount of carbon that can be taken into solid solution. For instance, at 1900 °F (1040 °C), approximately 1.6% C can be dissolved (Fig. 4).

Transformation of Austenite

Thus far the discussion has been confined to heating of the steel and the phases that result from various combinations of temperature and carbon content. Now what happens when the alloy is cooled? Referring to Fig. 4, assume that a steel containing 0.50% C is heated to 1500 °F (815 °C). All of the carbon will be dissolved (assuming of course that holding time is sufficient). Under these conditions, all of the carbon atoms will dissolve in the interstices of the fcc crystal (Fig. 2b). If the alloy is cooled slowly, transformation to the bcc (Fig. 2a) or alpha phase begins when the temperature drops below approximately 1450 °F (790 °C). As the temperature continues to decrease, the transformation is essentially complete at 1340 °F (725 °C). During this transformation, the carbon atoms escape from the lattice because they are essentially insoluble in the alpha crystal (bcc). Thus, in slow cooling, the alloy for all practical purposes, returns to the same state (in terms of phases) that it was before heating to form austenite. The same mechanism occurs with higher carbon steels, except that the austenite-to-ferrite transformation does not go through a two-phase zone (Fig. 4). In addition to the entry and exit of the carbon atoms through the interstices of the iron atoms, other changes occur that affect the practical aspects of heat treating.

First, a magnetic change occurs at 1418 °F (770 °C) as shown in Fig. 1. The heat of transformation effects many chemical changes, such as the heat that is evolved when water freezes into ice and the heat that is absorbed when ice melts. When an iron-carbon alloy is converted to austenite by heat, a large absorption of heat occurs at the transformation temperature. Likewise, when the alloy changes from gamma to alpha (austenite to ferrite), heat evolves.

What happens when the alloy is cooled rapidly? When the alloy is cooled suddenly, the carbon atoms cannot make an orderly escape from the iron lattice. This causes "atomic bedlam" and results in distortion of the lattice which manifests itself in the form of hardness and/or strength. If cooling is fast enough, a new structure known as martensite if formed, although this new structure (an aggregate of iron and cementite) is in the alpha phase.

Classification of Steels by Carbon Content

It must be remembered that there are only three phases in steels, but there are many different structures. A precise definition of eutectoid carbon is unavoidable; it varies according to the reference used. Figure 4 defines eutectoid carbon as 0.80%, but more recent investigations have shown that it is 0.77%. On the other hand, some phase diagrams show that the eutectoid is slightly higher than 0.80%. However, for the objectives of this book, the precise amount of carbon denoted as eutectoid is of no particular significance.

Hypoeutectoid Steels. Carbon steels containing less than 0.80% C are known as hypoeutectoid steels. The area bounded by AGB on the iron-cementite diagram (Fig. 4) is of significance to the room temperature microstructures of these steels; within the area, ferrite and austenite each having different carbon contents, can exist simultaneously.

Assume that a 0.40% carbon steel has been slowly heated until its temperature throughout the piece is 1600 °F (870 °C), thereby ensuring a fully austenitic structure. Upon slow cooling, free ferrite begins to form from the austenite when the temperature drops across the line AG, into the area AGB, with increasing amounts of ferrite forming as the temperature continues to decline while in this area. Ideally, under very slow cooling conditions, all of the free ferrite separates from austenite by the time the temperature of the steel reaches A_1 (the line BG) at 1340 °F (725 °C). The austenite islands, which remain at about 1340 °F (725 °C), now have the same amount of carbon as the eutectoid steel, or about 0.80%. At or slightly below 1340 °F (725 °C), the remaining untransformed austenite transforms—it becomes pearlite which is so named because of its resemblance to mother of pearl. Upon further cooling to room temperature, the microstructure remains unchanged, resulting in a final room temperature microstructure of an intimate mixture of free ferrite grains and pearlite islands.

A typical microstructure of a 0.40% carbon steel is shown in Fig. 5(a). The pure white areas are the islands of free ferrite grains described above. Grains that are white but containing dark platelets are typical lamellar pearlite. These platelets are cementite or carbide interspersed through the ferrite, thus conforming to the typical two-phase structure indicated below the BH line in Fig. 4.

Eutectoid Steels. A carbon steel containing approximately 0.77% C becomes a solid solution at any temperature in the austenite temperature range, i.e., between 1340 and 2500 °F (725 and 1370 °C). All of the carbon is dissolved in the austenite. When this solid solution is slowly cooled, several changes occur at 1340 °F (725 °C). This temperature is a transformation temperature or critical temperature of the iron-cementite system. At this temperature, a 0.77% (0.80%) carbon steel transforms from a single homogeneous solid solution into two distinct new solid phases. This change occurs at constant temperature and with the evolution of heat. The new phases are ferrite and cementite, formed simultaneously; however, it is only at composition point G in Fig. 4 (0.77% carbon steel) that this phenomenon of the simultaneous formation of ferrite and cementite can occur.

The microstructure of a typical eutectoid steel is shown in Fig. 5(b). This represents a polished and slightly etched specimen at a magnification of 750X. The white matrix is alpha ferrite and the dark platelets are cementite. All grains are pearlite—no free ferrite grains are present under these conditions.

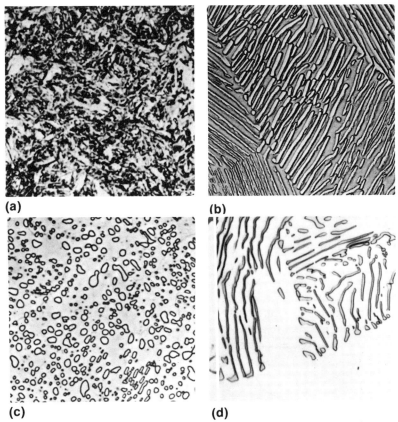

Fig. 5 Microstructures of a 0.40% carbon steel. (a) Ferrite grains (white) and pearlite (gray streaks) in a white matrix of a hypoeutectoid steel. 1000×. (b) Microstructure of a eutectoid steel (all pearlite grains). 2000×. (c) Microstructure of a eutectoid steel with all cementite in the spheroidal form. 1000×. (d) Microstructure of a hypereutectoid steel—pearlite with excess cementite bounding the grains, see text. 1000×. Source: (a), (b), (c), *Metals Handbook*, Vol 7, 8th edition, American Society for Metals, 1972. (d) MEI Course 1, Lesson 11, American Society for Metals, p 11, 1977.

Cooling conditions (rate and temperature) govern the final condition of the particles of cementite that precipitate from the austenite at 1340 °F (725 °C). Under specific cooling conditions, the particles become spheroidal instead of elongated platelets as shown in Fig. 5(b). Figure 5(c) shows a similar two-phase structure resulting from slowly cooling a eutectoid carbon steel just below A_1. This structure is commonly known as spheroidite, but is still a dispersion of cementite particles in alpha ferrite. There is no indication of grain boundaries in Fig. 5(c). The spheroidized structure is often preferred over the pearlitic structure because spheroidite has superior machinability and formability. Combination structures (that is, partly lamellar and partly spheroidal cementite in a ferrite matrix) are also common.

As noted above, a eutectoid steel theoretically contains a precise amount of carbon. In practice, steels that contain carbon within the range of approximately 0.75 to 0.85% are commonly referred to as eutectoid carbon steels.

Hypereutectoid steels contain carbon contents of approximately 0.80 to 2.0%. Assume that a steel containing 1.0% C has been heated to 1550 °F (845 °C), thereby ensuring a 100% austenitic structure. When cooled, no change occurs until the line GF (Fig. 4), known as the A_{cm} or cementite solubility line, is reached. At this point, cementite begins to separate out from the austenite, and increasing amounts of cementite separate out as the temperature of the 1% carbon steel descends below the A line. The composition of austenite changes from 1% C toward 0.77% C. At a temperature slightly below 1340 °F (725 °C), the remaining austenite changes to pearlite. No further changes occur as cooling proceeds toward room temperature, so that the room temperature microstructure consists of pearlite and free cementite. In this case, the free cementite exists as a network around the pearlite grains (Fig. 5d).

Upon heating hypereutectoid steels, reverse changes occur. At 1340 °F (725 °C), pearlite changes to austenite. As the temperature increases above 1340 °F (725 °C), free cementite dissolves in the austenite, so that when the temperature reaches the A_{cm} line, all the cementite dissolves to form 100% austenite.

Hysteresis in Heating and Cooling

The critical temperatures (A_1, A_s, and A_{cm}) are "arrests" in heating or cooling and have been symbolized with the letter A, from the French word "arret" meaning arrest or a delay point, in curves plotted to show heating or cooling of samples. Such changes occur at transformation temperatures in the iron-cementite diagram if sufficient time is given and can be plotted for steels showing lags at transformation temperatures, as shown for iron in Fig. 4. However, because heating rates in commercial practice usually exceed those in controlled laboratory experiments, changes on heating usually occur at temperatures a few degrees above the transformation temperatures shown in Fig. 4 and are known as Ac temperatures, such as Ac_1 or Ac_3. The "c" is from the French word "chauffage," meaning heating. Thus, Ac_1 is a few degrees above the ideal A_1 temperature.

Likewise, on slow cooling in commercial practice, transformation changes occur at temperatures a few degrees below those on Fig. 4. These are known as Ar, or Ar_3, the "r" originating from the French word "refroidissement," meaning cooling.

This difference between the heating and cooling varies with the rate of heating or cooling. The faster the heating, the higher the Ac point; the faster the cooling, the lower the Ar point. Also, the faster the heating and cooling rate, the greater the gap between the Ac and Ar points of the reversible (equilibrium) point A.

Going one step farther, in cooling a piece of steel, it is of utmost importance to note that the cooling rate may be so rapid (as in quenching steel in water) as to suppress the transformation for several hundreds of degrees. This is due to the decrease in reaction rate with decrease in temperature. As discussed below, time is an important factor in transformation, especially in cooling.

Effect of Time on Transformation

The foregoing discussion has been confined principally to phases that are formed by various combinations of composition and temperature; little reference has been made to the effects of time. In order to convey to the reader the effects of time on transformation, the simplest approach is by means of a TTT curve (time, temperature, transformation) for some constant iron-carbon composition.

Such a curve is presented in Fig. 6 for a 0.77% (eutectoid) carbon steel. TTT curves are also known as "S" curves because the principal part of the curve is shaped like the letter "S." In Fig. 6, temperature is plotted on the vertical axis, and time is plotted on a logarithmic scale along the horizontal axis. The reason for plotting time on a logarithmic scale is merely to keep the width of the chart within a manageable dimension.

In analyzing Fig. 6, begin with line Ae$_1$ (1340 °F or 725 °C). Above this temperature, austenite exists only for a eutectoid steel (refer also to Fig. 4). When the steel is cooled and held at a temperature just below Ae$_1$ (1300 °F or 705 °C) transformation begins (follow line P$_s$, B$_s$), but very slowly at this temperature; 1 h of cooling is required before any significant amount of transformation occurs, although eventually complete transformation occurs isothermally (meaning at a constant temperature), and the transformation product is spheroidite (Fig. 5c). Now assume a lower temperature (1200 °F or 650 °C) on line P$_s$ B$_s$ (the line of beginning transformation); transformation begins in less than 1 min, and the transformation product is coarse pearlite (near the right side of Fig. 6). Next assume a temperature of 1000 °F (540 °C); transformation begins in approximately 1 s and is completely transformed to fine pearlite in a matter of a few minutes. The line Pf B$_f$ represents the completion of transformation and is generally parallel with Ps B$_s$. However, if the steel is cooled very rapidly (such as by immersing in water) so that there is not sufficient time for transformation to begin in the 1000 °F (540 °C) temperature vicinity, then the beginning of transformation time is substantially extended. For example, if the steel is cooled to and held at 600

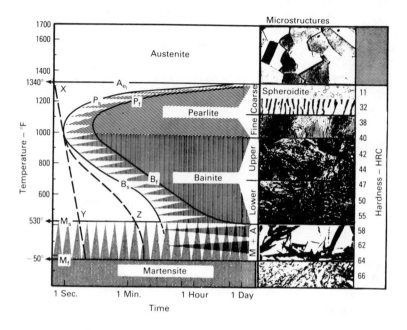

Fig. 6 Time-temperature-transformation (TTT) diagram for a eutectoid (0.77%) carbon steel. Source: MEI Course 10, Lesson 3, American Society for Metals, p 5, 1981.

°F (315 °C), transformation does not begin for well over 1 min. It must be remembered that all of the white area to the left of line Ps Bs represents the austenitic phase, although it is highly unstable. When transformation takes place isothermally within the temperature range of approximately 550 to 800 °F (290 to 425 °C), the transformation product is a microstructure called bainite (upper or lower as indicated toward the right of Fig. 6). A bainitic microstructure is shown in Fig. 7(b). In another example, steel is cooled so rapidly that no transformation takes place in the 1000 °F (540 °C) region and rapid cooling is continued (note line XY in Fig. 6) to and below 530 °F (275 °C) or Ms. Under these conditions, martensite is formed. Point Ms is the temperature at which martensite begins to form, and Mf indicates the complete finish of transformation. It must be remembered that martensite is not a phase, but is a specific microstructure in the ferritic (alpha) phase. Martensite is formed from the carbon atoms jamming the lattice of the austenitic atomic arrangement. Thus, martensite can be considered as an aggregate of iron and cementite (Fig. 7a).

In Fig. 6, the microstructure of austenite (as it apparently appears at elevated temperature) is shown on the right. It is also evident that the lower the temperature at which transformation takes place, the higher the hardness (see Chapter 3).

It is also evident that all structures from the top to the region where martensite forms (Ae_1) are time-dependent, but the formation of martensite is not time-dependent.

Each different steel composition has its own TTT curve; Fig. 6 is presented only as an example. However, patterns are much the same for all steels as far as shape of the curves is concerned. The most outstanding difference in the curves among different steels is the distance between the vertical axis and the nose of the S curve. This occurs at about 1000 ° F (540 ° C) for the steel in Fig. 6. This distance in terms of time is about 1 s for a eutectoid carbon steel, but could be an hour or more for certain high-alloy steels, which are extremely sluggish in transformation.

The distance between the vertical axis and the nose of the S curve is often called the "gap" and has a profound effect on how the steel must be cooled to form the hardened structure—martensite. Width of this gap for any steel is directly related to the critical cooling rate for that specific steel. Critical cooling rate is defined as the rate at which a given steel must be cooled from the austenite to prevent the formation of nonmartensitic products.

In Fig. 6, it is irrelevant whether the cooling rate follows the lines X to Y or X to Z because they are both at the left of the beginning transformation line Ps B_s. Practical heat treating procedures are based on the fact that once the steel has been cooled below approximately 800 ° F (425 ° C), the rate of cooling may be decreased. The conditions as described above are all closely related to hardenability which will be dealt with in Chapter 3.

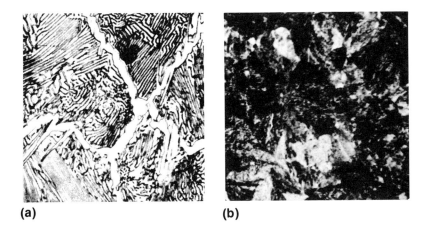

(a)   **(b)**

Fig. 7 (a) Microstructure of quenched 0.95% carbon steel. 1000×. Structure is martensitic. (b) Bainitic structure in a quenched 0.95% carbon steel. 550×. Source: *Metals Handbook,* Vol 7, 8th edition, American Society for Metals, 1972.

Chapter 3

Hardness and Hardenability

Chapter 2 dealt exclusively with the fundamentals involved in the heat treatment of steel, which included the basic mechanism of hardening. However, when one begins to apply this fundamental knowledge to the hardening of steel, several oddities immediately become evident. First, it may be observed that one steel becomes a great deal harder than another, in spite of the fact that both have been hardened according to the correct principles. Second, it may be observed that, when two steels are hardened, one may be as hard in the center of a bar as it is on the surface, but another may be a great deal softer in the center than it is at the surface. This occurs in spite of the fact that both bars are identical in diameter and have been correctly processed. Naturally, one wants to know why such differences exist. In order to explain this, it is necessary to obtain and understand an accurate and reliable measuring device.

There are several methods for measuring or evaluating the results attained from heat treating—tension testing, impact testing, bend testing, and the various types of shear testing, to name a few. However, as a rule, the types of tests mentioned above are restricted to special applications. The hardness test is, by far, the most universally used device for measuring the results of heat treating for several reasons—(1) the hardness test is simple to perform; (2) it does not usually impair the usefulness of the workpiece being tested; and (3) accurately made and properly interpreted hardness tests can be used to evaluate (or at least estimate) other mechanical properties. For example, hardness varies directly with strength and inversely with ductility.

What Is Hardness?

Since childhood, we have been acquainted with hardness, in one way or another, and we assume that we are well informed on so simple a matter. If, however, we should ask our neighbors for a definition of hardness, the results would be quite surprising. To some, a hard substance is one that resists wear; to others, something that does not bend; to someone else, an object that resists penetration; and to still others, an object that breaks or scratches other substances. From the variety of answers given, it is apparent that there are many different opinions about what actually constitutes hardness. Thus, if an attempt is made to evaluate the answers and arrive at some common definition, it becomes apparent that hardness is an elusive property, far more complex than most people would believe.

The definition provided in most dictionaries is "the relative capacity of a substance for scratching another or for being scratched or indented by another." While this definition is somewhat vague, it is probably the best sense of the term that can be conveyed in a few words. For example, if steel and glass are considered with respect to the various definitions above, the difficulty of evaluating hardness is better appreciated. For example, both glass and steel have been used as paving materials in sidewalks, and their resistance to wear has been observed on many occasions; from the viewpoint of pure wear, it is known that both endure equally well. Therefore, it might be said that one is as hard as the other. It is true, moreover, that glass actually can scratch steel, which would appear to place glass in a harder category. Steel, however, can crush or break glass, and this would appear to make it harder than glass. The hopelessness of trying to evaluate hardness on the basis of numerous properties, without a concise definition of what property we want to measure as hardness, is readily apparent.

The difficulty of defining and measuring hardness was recognized early in recorded history and continues to exist today. Any attempt to make an all-inclusive definition leads only to confusion. For this reason, in both ferrous and nonferrous metallurgy, hardness is defined and measured on the basis of (1) a static test in which a standard penetrator is pressed into the specimen, using a standard load, and the resistance of the metal to penetration is the measure of the hardness; and (2) a dynamic test in which a small hammer is dropped against the surface of the material from a fixed height and the hardness expressed in terms of height to which the hammer rebounds.

History of Hardness Testing

According to the records, the first method of evaluating hardness was developed in 1772, when Reaumur devised a procedure of pressing the edges of two prisms together, as shown in Fig. 1(a). The pressure applied to each prism was the same and thus allowed a direct comparison of hardness between the two metals.

In 1897, Foeppl altered the original work by Reaumur by pressing together two semicylindrical bars instead of two prisms, as shown in Fig. 1(b). By measuring the area of contact of the flattened surfaces and dividing this area into the load, he obtained a hardness evaluation (Ref. 5).

A few years later, Brinell introduced his well-known method, consisting of

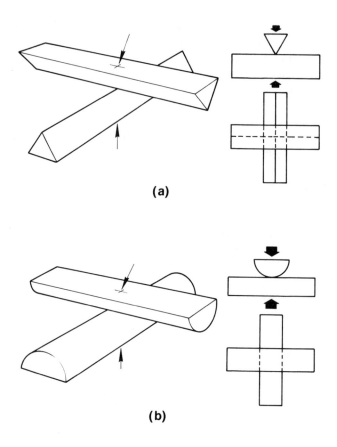

(a)

(b)

Fig. 1 Early attempts at evaluating hardness of metals. (a) Reaumur's method of determining comparative hardness. Edges of two metal prism specimens are pressed together. (b) Focppl's method of hardness testing. Two semicylindrical bars are pressed together and the flattened area of contact is measured. This technique is also used in hot hardness testing. Source: MEI Course 10, Lesson 6, American Society for Metals, p 2, 1977.

pressing a hard steel ball into the surface to be tested, determining the surface area of the impression made, and dividing this value into the load, thus arriving at a hardness value. This test still is extensively used and represents the first hardness test that yields reproducible results. Numerous modifications of hardness tests have been developed that utilize a wide variety of penetrators and loads imposed on the penetrators (indenters). In most instances, hardness was evaluated by using a constant load, then determining either the depth or projected area of the indentation. The reverse has also been used; that is, forcing the penetrator to a pre-established depth, then measuring the required load. Each of the variations has fields of usefulness that are quite valuable and illustrate that, even though it is possible to define hardness (at least as applied to metals), a single universal testing procedure is not feasible.

Hardness Testing Systems

Many different systems have been devised for testing the hardness of metals, but only a few have achieved commercial importance. They are the Brinell, Rockwell, Vickers, Scleroscope, and various microhardness testers that employ Vickers or Knoop indenters. All of these systems, except the Scleroscope, depend strictly on the principle of indentation for measurement.

For the reader to have a clear understanding of heat treating, it is essential that he understand the commonly used hardness testing systems. Therefore, each of the above-mentioned systems is discussed below. For detailed descriptions of these systems and their application, see Ref 6.

Brinell Testing. The Brinell test is simple and consists of applying a constant load, usually 500 to 3000 kg, or a hardened steel ball type of indenter, 10 mm in diameter, to the flat surface of a workpiece (Fig. 2). The 500-kg load usually is used for testing nonferrous metals such as copper and aluminum alloys, whereas the 3000-kg load is most often used for testing harder metals such as steels and cast irons. The load is held for a specified time (10 to 15 s for iron and steel and about 30 s for softer metals), after which the diameter of the recovered indentation is measured in millimeters. This time period is required to ensure that plastic flow of the work metals has stopped.

Hardness is evaluated by taking the mean diameter of the indentation (two readings at right angles to each other) and calculating the Brinell hardness number (HB) by dividing the applied load to the surface area of the indentation according to the following formula:

$$HB = \frac{L}{\frac{\pi D}{2}(D - \sqrt{D^2 - d^2})}$$

Where L is the load, in kilograms; D is the diameter of the ball, in millimeters; and d is the diameter of the indentation, in millimeters. It is not necessary, however, to make the calculation for each test. Tables, such as Table 1 (for 500-, 1500-, and 3000-kg loads), are provided that give the hardness in Brinell numbers for each diameter of indentation. Highly hardened steels cannot be tested by the Brinell method because the steel ball indenter will deform. The practical upper limit for Brinell testing is an indentation of about 2.90 mm diameter of 444 HB. The Brinell indenter can be loaded by any one of a number of different machines. There are several machines designed for the purpose, although Brinell testing can be performed in a conventional hydraulic tension-compression testing machine. Disadvantages of the Brinell test (in addition to the upper limit of hardness) are (1) a great deal of time is required—at least 1 min for applying the load then reading the impression diameter by means of a specially designed microscope; (2) because the impression made by the indenter is relatively large, small workpieces cannot be tested; and (3) the impression left by such a large indenter may be damaging to the workpiece. To reduce the amount of time required for testing, however, high-speed, automatic, direct-reading machines now are available. Although readings are in Brinell numbers, indentation depth is measured instead of diameter.

Rockwell Testing. The Rockwell test is the most versatile of all hardness testers. By means of various applied loads and a number of different indenters, the Rockwell test can be used for testing all metals from soft solder to highly hardened steels and carbides.

Rockwell hardness testing differs from Brinell hardness testing in that hardness is determined by the depth of indentation made by a constant load impressed upon an indenter. Although a number of different indenters are used for Rockwell hardness testing, the most common type is a diamond, ground to a 120° cone with a spherical apex having a 0.2-mm radius, which is known as a Brale indenter.

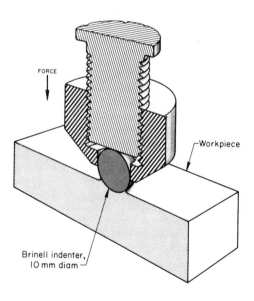

Fig. 2 Sectional view of a Brinell indenter, showing the manner in which the application of force by the indenter causes the metal of the workpiece to flow. Source: *Metals Handbook*, Vol 11, 8th edition, American Society for Metals, p 3, 1976.

Table 1. Brinell hardness numbers for indentation diameters of 2.45 to 6.45 mm obtained with three different loads on 10 mm diam. ball indenters.

Source: Metals Handbook, Vol 11, 8th edition, American Society for Metals, p 4, 1976.

Indentation diam, mm	Brinell hardness — Load, kg			Indentation diam, mm	Brinell hardness — Load, kg			Indentation diam, mm	Brinell hardness — Load, kg		
	500	1500	3000		500	1500	3000		500	1500	3000
2.45	104	314	627	3.80	42.4	128	255	5.15	22.3	67.0	134
2.50	100	301	601	3.85	41.3	124	248	5.20	21.8	65.5	131
2.55	96.3	289	578	3.90	40.2	121	241	5.25	21.4	64.0	128
2.60	92.6	278	555	3.95	39.1	118	235	5.30	20.9	63.0	126
2.65	89.0	267	534	4.00	38.1	115	229	5.35	20.5	61.5	123
2.70	85.7	257	514	4.05	37.1	112	223	5.40	20.1	60.5	121
2.75	82.6	248	495	4.10	36.2	109	217	5.45	19.7	59.0	118
2.80	79.6	239	477	4.15	35.3	106	212	5.50	19.3	58.0	116
2.85	76.8	231	461	4.20	34.4	104	207	5.55	18.9	57.0	114
2.90	74.1	222	444	4.25	33.6	101	201	5.60	18.6	55.5	111
2.95	71.5	215	429	4.30	32.8	98.5	197	5.65	18.2	54.5	109
3.00	69.1	208	415	4.35	32.0	96.0	192	5.70	17.8	53.5	107
3.05	66.8	201	401	4.40	31.2	93.5	187	5.75	17.5	52.5	105
3.10	64.6	194	388	4.45	30.5	91.5	183	5.80	17.2	51.5	103
3.15	62.5	188	375	4.50	29.8	89.5	179	5.85	16.8	50.5	101
3.20	60.5	182	363	4.55	29.1	87.0	174	5.90	16.5	49.6	99.2
3.25	58.6	176	352	4.60	28.4	85.0	170	5.95	16.2	48.7	97.3
3.30	56.8	171	341	4.65	27.8	83.5	167	6.00	15.9	47.8	95.5
3.35	55.1	166	331	4.70	27.1	81.5	163	6.05	15.6	46.9	93.7
3.40	53.4	161	321	4.75	26.5	79.5	159	6.10	15.3	46.0	92.0
3.45	51.8	156	311	4.80	25.9	78.0	156	6.15	15.1	45.2	90.3
3.50	50.3	151	302	4.85	25.4	76.0	152	6.20	14.8	44.4	88.7
3.55	48.9	147	293	4.90	24.8	74.5	149	6.25	14.5	43.6	87.1
3.60	47.5	143	285	4.95	24.3	73.0	146	6.30	14.2	42.8	85.5
3.65	46.1	139	277	5.00	23.8	71.5	143	6.35	14.0	42.0	84.0
3.70	44.9	135	269	5.05	23.3	70.0	140	6.40	13.7	41.3	82.5
3.75	43.6	131	262	5.10	22.8	68.5	137	6.45	13.5	40.5	81.0

As shown in Fig. 3(a), the Rockwell hardness test consists of measuring the additional depth to which an indenter is forced by a heavy (major) load (Fig. 3b) beyond the depth of a previously applied light (minor) load (Fig. 3a). Application of the minor load eliminates backlash in the load train and causes the indenter to break through slight surface roughness and to crush particles of foreign matter, thus contributing to much greater accuracy in the test. The basic principle involving minor and major loads illustrated in Fig. 3 applies to steel ball indenters as well as to diamond indenters.

The minor load is applied first, and a reference or "set" position is established on the dial gage of the Rockwell hardness tester. Then the major load is applied. Without moving the workpiece being tested, the major load is removed, and the Rockwell hardness number is automatically indicated on the dial gage. The entire operation takes 5 to 10 s.

Diamond indenters are used mainly for testing material such as hardened steels and cemented carbides. Steel ball indenters, available with diameters of 1/16, 1/8, 1/4, and 1/2 in., are used when testing materials such as soft steel, copper alloys, aluminum alloys, and bearing metals. Each load and indenter combination has a specified letter designation as indicated in Table 2. Regardless of the scale, hardness readings are written with the number first (dial indication); the letters HR (hardness Rockwell) follow, and finally the scale designation. For example, 60 Rockwell "C" (diamond indenter with a 150-kg load) is written as "60 HRC." Likewise, a reading of 90 Rockwell "B" would be written as "90 HRB." The Rockwell "C" scale was developed first and is still the most widely used regardless of whether the test is actually conducted with a diamond indenter and a 150-kg load, or whether the actual test is made with another scale and converted to the "C" scale (conversion tables are readily available, see Ref 6). When hardness readings are reported,

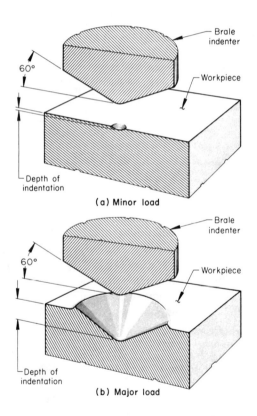

Fig. 3 Indentation in a workpiece made by application of (a) the minor load and (b) the major load, on a diamond Brale indenter in Rockwell hardness testing. The hardness value is based on the difference in depths of indentation produced by the minor and major loads. Source: *Metals Handbook*, Vol 11, 8th edition, p 6, 1976.

it is essential that it be indicated where conversions have been made to obtain the listed values.

Rockwell testers vary widely in design. Figure 4 shows the components of a common type of Rockwell tester. Regardless of tester design, the principles are the same. Consequently, testers are now available in an almost infinite number of designs to accommodate a large variety of testing requirements. Models are available that are completely automated, have digital readouts, and other marks of sophistication.

Vickers Testing. The Vickers hardness test is similar to the Brinell principle in that an indenter of definite shape is pressed into the material to be tested, the load removed, the diagonals of the resulting indentation are measured, and the hardness number is calculated based on the area of indentation. The Vickers indenter is a square-based pyramid that has an angle of 136° between faces, as shown in Fig. 5.

With the Vickers indenter, the depth of indentation is about one seventh of the diagonal length of the indentation. The Vickers hardness number (HV) is the ratio of the load applied to the indenter to the surface area of the indentation. However, as is the case for Brinell testing, tables are posted on testers that provide the Vickers hardness number based on load and the diagonal readings made with a microscope—an integral part of the Vickers tester.

The range of hardness that can be accommodated by the Vickers test method is quite broad because of the various loads that can be used. The Vickers test is, however, less versatile than the Rockwell test and more time-consuming. In a Vickers tester, the load range is 1 to 120 kg; however, for

Table 2. Rockwell hardness scale designations for types of indenters and loads.
Source: Metals Handbook, Vol 11, 8th edition, American Society for Metals, p 7, 1976.

Scale desig- nation	Indenter Type	Diam. in.	Major load, kg	Dial figure	Scale desig- nation	Indenter Type	Diam. in.	Major load, kg
Regular Rockwell Tester					**Superficial Rockwell Tester**			
B	Ball	$\frac{1}{16}$	100	Red	15N	N Brale	..	15
C	Brale	..	150	Black	30N	N Brale	..	30
A	Brale	..	60	Black	45N	N Brale	..	45
D	Brale	..	100	Black	15T	Ball	$\frac{1}{16}$	15
E	Ball	$\frac{1}{8}$	100	Red	30T	Ball	$\frac{1}{16}$	30
F	Ball	$\frac{1}{16}$	60	Red	45T	Ball	$\frac{1}{16}$	45
G	Ball	$\frac{1}{16}$	150	Red	15W	Ball	$\frac{1}{8}$	15
H	Ball	$\frac{1}{8}$	60	Red	30W	Ball	$\frac{1}{8}$	30
K	Ball	$\frac{1}{8}$	150	Red	45W	Ball	$\frac{1}{8}$	45
L	Ball	$\frac{1}{4}$	60	Red	15X	Ball	$\frac{1}{4}$	15
M	Ball	$\frac{1}{4}$	100	Red	30X	Ball	$\frac{1}{4}$	30
P	Ball	$\frac{1}{4}$	150	Red	45X	Ball	$\frac{1}{4}$	45
R	Ball	$\frac{1}{2}$	60	Red	15Y	Ball	$\frac{1}{2}$	15
S	Ball	$\frac{1}{2}$	100	Red	30Y	Ball	$\frac{1}{2}$	30
V	Ball	$\frac{1}{2}$	150	Red	45Y	Ball	$\frac{1}{2}$	45

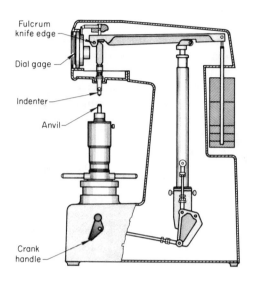

Fig. 4 Principal components of a regular (normal) Rockwell hardness tester. Source: *Metals Handbook*, Vol 11, 8th edition, American Society for Metals, p 6, 1976.

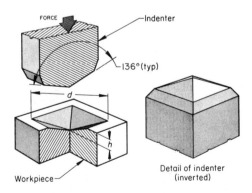

Fig. 5 Schematic representation of the square-based pyramidal diamond indenter used in a Vickers hardness tester and of the resulting indentation in the workpiece. Source: *Metals Handbook*, Vol 11, 8th edition, American Society for Metals, p 13, 1976.

most hardness testing, 50 kg is maximum. The Vickers test also can be used for microhardness testing as described below.

Scleroscope Testing. The Scleroscope hardness test is essentially a dynamic indentation test wherein a diamond-tipped hammer is dropped from a fixed height onto the surface of the material being tested. The height of rebound of the hammer is a measure of the hardness of the material. The Scleroscope scale consists of units that are determined by dividing the average rebound of the hammer from a quenched (to maximum hardness) and untempered water-hardening tool steel into 100 units. The scale is continued above 100 to permit testing of materials having hardness greater than that of fully hardened tool steel. Scleroscope hardness testing can be conducted rapidly, and some testing instruments are portable so that they can be used for testing large workpieces that would be difficult to bring to the tester.

Microhardness Testing

Microhardness testing is usually done with loads that do not exceed 1 kg and may be as little as 1 g, although the most common load range is 100 to 500 g. In general, however, the term "microhardness" relates to the size of the indentation rather than to the applied load.

Microhardness testing is capable of providing information on hardness characteristics that cannot be provided by the more conventional methods. It is not, however, strictly a research tool and is used extensively for control of many production operations. Specific fields of application include:

• Measuring hardness of precision workpieces that are too small to be measured by other methods
• Measuring hardness of product forms such as foils that are too thin to be measured by other methods
• Monitoring of carburizing or nitriding operations, which usually is accomplished by hardness surveys taken on cross sections of test pieces that accompanied the workpieces through production operations
• Measuring hardness of individual microconstituents
• Measuring hardness close to edges, detecting undesirable surface conditions such as grinding burn and decarburization
• Measuring hardness of surface layers such as plating or bonded layers

Microhardness testing can be performed with either the Vickers indenter (Fig. 5) or the Knoop indenter shown in Fig. 6.

Knoop indentation testing is performed with a diamond, ground to pyramidal form, that produces a diamond-shaped indentation having an approximate ratio between long and short diagonals of 7 to 1. The pyramid shape employed has an included longitudinal angle of 172° 30' and an included transverse angle of 130°. The depth of indentation is about 1/30th of its length. Because of the shape of the indenter, indentations of accurately measurable length are obtained with light loads.

The Knoop hardness number (HK) is the ratio of the load applied to the indenter to the unrecovered projected area of indentation. By formula:

$$HK = \frac{P}{A} = \frac{P}{CL^2}$$

Where P is the applied load, in kilograms; A is the unrecovered projected area of indentation, in square millimeters; and C is 0.07028, a constant of the indenter relating projected area of the indentation to the square of the length of the long diagonal.

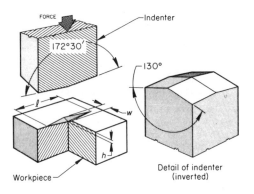

Fig. 6 Schematic representation of a pyramidal Knoop indenter and of the resulting indentation in the workpiece. Source: *Metals Handbook*, Vol 11, 8th edition, p 16, 1976.

The above formula is, however, mainly for academic interest, as tables are provided that provide the Knoop hardness for the size of any indentation (measured in microns) and the applied load. Superiority of the Vickers versus the Knoop indenter is often controversial. One advantage of the Knoop indenter is that only the length of the indentation is measured (Fig. 7), whereas for greatest accuracy, both diagonals of the Vickers indentation should be taken and then averaged.

Figure 7 presents a comparison of the indentations made by Knoop and Vickers indenters. Each has some advantages over the other. For example, the Vickers indenter penetrates about twice as far into the workpiece as does the Knoop indentation. Therefore, the Vickers indenter is less sensitive to minute differences in surface condition than the Knoop indenter. However, the Vickers indentation, because of the shorter diagonal, is more sensitive to errors in measuring than the Knoop indentation.

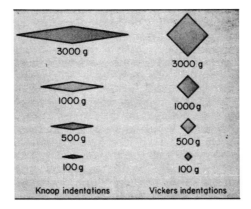

Fig. 7 Comparison of indentations made by Knoop and Vickers indenters in the same work metal and at the same loads. Source: *Metals Handbook*, Vol 11, 8th edition, American Society for Metals, p 16, 1976.

Several types of microhardness testers are available. The most accurate operate through the direct application of load by dead weight, or by weights and levers. Principal components of a typical microhardness tester are shown in Fig. 8, which demonstrates how the indentation is made on the specimen, as well as the repositioning of the specimen for measurement by a special type of microscope.

Effect of Carbon Content on Annealed Steels

In Chapter 2, two metallurgical phenomena—the allotropy of iron and the small size of the carbon atom—were described as key factors in the heat treatment of steel. Chapter 2 also discussed how carbon atoms exit from the face-centered cubic (fcc) iron crystal in a slow and orderly manner, resulting in an annealed (soft) condition. Rapid cooling entraps the carbon atoms in the lattice; consequently, the structure changes from an fcc to a body-centered cubic (bcc) lattice structure, as high strength and hardness result. Also, referring to the TTT diagram (Fig. 6) in Chapter 2, note that the hardness of martensite (the fully hardened structure) is 65 HRC.

Logically, the reader might wonder at this point what affect increasing carbon content has on annealed steels. To be sure, there is an effect. Referring to the iron-cementite phase diagram (Fig. 4) in Chapter 2, near the extreme left—at approximately 0.20% C, about one-fourth of the grains are pearlite, while the remainder is nearly pure ferrite (see microstructures associated with the iron-iron carbide diagram, Fig. 5). The ferrite grains are very soft (below the HRC scale), while the pearlite grains are considerably harder because they contain a dispersion of cementite in the ferrite. Moving to right of the iron-

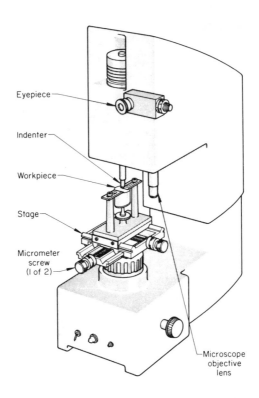

Fig. 8 Principal components of a typical microhardness tester. Source: *Metals Handbook*, Vol 11, 8th edition, American Society for Metals, p 16, 1976.

cementite diagram, the number of pearlite grains increases in proportion to the number of ferrite grains until the eutectoid (0.77% C) is reached. At this point, 100% pearlite grains exist. Thus, hardness, as measured by indentation methods, increases as the ratio of pearlite-to-ferrite grains increases, because the cementite particles act as reinforcement to the ferrite and resist being "pushed" by an indenter. This condition generally exists even when the cementite particles are spheroidal (see Fig. 5 in Chapter 2). However, the spheroidal particles offer less resistance to force from the indenter compared with a steel of the same carbon content with a structure of lamellar pearlite. Therefore, the spheroidal structure registers as the softer of the two materials.

Now, let us go a step further and assume that a steel containing 1.0% C or higher (a hypereutectoid steel) is cooled slowly so that lamellar pearlite is formed. With the carbon content higher than a eutectoid alloy, there is an excess of cementite that often gathers at the grain boundaries as shown in Fig.

5(d) in Chapter 2. This cementite is relatively hard, thus offering increased resistance to indentation. Therefore, hardness does increase as carbon content increases, but not dramatically for annealed steels; that is, not as measured by a conventional hardness tester. Likewise, yield strength and tensile strength increase in annealed carbon steels as carbon content increases.

It should be noted at this point that there are two types of hardness—apparent and true. Apparent hardness is the actual reading taken from an instrument such as a Brinell or Rockwell tester. When a relatively large indenter is forced into the metal specimen, the indenter can register total resistance of the mass into which it is being forced; when two or more microconstituents are involved, the result is an average of these values.

Conversely, true hardness involves testing with a microhardness tester so as to obtain the true hardness values of the various microconstituents. For example, cementite is far harder than ferrite. This accounts for the fact that hypereutectoid steels have far greater resistance to abrasive wear than hypoeutectoid or eutectoid steels, even though all steels are in the annealed condition. The difference in apparent hardness is not great. Actually, the tiny particles of cementite embedded in a matrix act like sand embedded in a plastic surface.

Role of Carbon in Hardened Steels

As emphasized in Chapter 2, carbon is the key to the hardening of steels by the heating and quick cooling (quenching) mechanism. The carbon content of a steel determines the maximum hardness attainable. The effect of carbon on attainable hardness is demonstrated in Fig. 9. The maximum attainable hardness requires only about 0.60% C, which probably seems odd to the reader. From the iron-cementite diagram (Fig. 4) in Chapter 2, it would seem logical that hardness would not become a straight line until about 0.77% C is reached. However, there is essentially no change in attainable hardness above about 0.60% C according to the data shown in Fig. 9. This can be accounted for through some unavoidable deficiencies concerned with indentation hardness testing. Despite this apparent oddity concerning Fig. 9, these data are accurate and absolutely reproducible for extremely thin sections of carbon steel.

In fact, the data shown in Fig. 9 are precise to the extent that they sometimes are used in reverse; that is, they can be used as a quick means of determining carbon content of an unknown steel. For example, when the carbon content of a steel is unknown and analytical means are not available, cutting wafer-thin sections of the unknown steel, heating them above the transformation temperature, and quenching them in water helps aid in identification of the material by observing and measuring the physical and

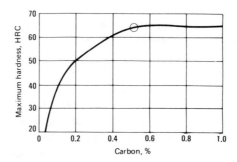

Fig. 9 Relationship between carbon content and maximum hardness. Full hardness can be obtained with as little as 0.60% C. As noted by data point. Source: *Heat Treaters' Guide: Standard Practices and Procedures for Steel*, American Society for Metals, p 20, 1982.

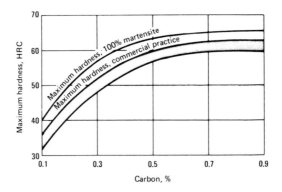

Fig. 10 Relationship between carbon content and maximum hardness. Usually attained in commercial hardening. Source: *Heat Treaters' Guide: Standard Practices and Procedures for Steel*, American Society for Metals, p 20, 1982.

mechanical properties that result. The specimens then are measured for hardness; if specimens show hardness values of approximately 50 HRC, the carbon content is about 0.20%; for readings of around 60 HRC, the carbon content is about 0.40%. Obviously, as shown in Fig. 9, this method of carbon determination would not be valid much above 60 HRC because the hardness line starts to become a straight line.

In conventional heat treating practice, it should be remembered that the data shown in Fig. 9 are theoretical; data are based on heat treating of wafer-thin sections that are cooled from their austenitizing temperature to room temperature within a matter of seconds, thus developing 100% martensite throughout their sections. Therefore, the ideal condition shown in Fig. 9 is seldom attained in practice. Figure 10 shows a better example of hardness versus carbon content, because it is a more accurate condition, one expected in commercial practice.

The most important factor influencing the maximum hardness that can be attained is mass of the metal being quenched. In a small section, the heat is extracted quickly, thus exceeding the critical cooling rate of the specific steel. The critical cooling rate is the rate of cooling that must be exceeded to prevent formation of nonmartensite products. As section size increases, it becomes increasingly difficult to extract the heat fast enough to exceed the critical cooling rate and thus avoid formation of nonmartensitic products. A typical condition is shown in Fig. 11, which illustrates the effect of section size on surface hardness and is a good example of the mass effect. For small sections up to 0.5 in. (13 mm), full hardness of about 63 or 65 HRC is attainable. As the diameter of the quenched piece is increased, cooling rates and hardness decrease, because the critical cooling rate for this specific steel is not exceeded.

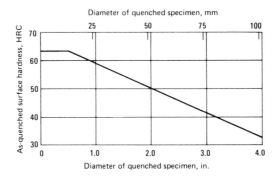

Fig. 11 Effect of section size on surface hardness of an 0.54% carbon steel. Quenched in water from 1525 °F (830 °C). Source: *Heat Treaters' Guide: Standard Practices and Procedures for Steel*, American Society for Metals, p 21, 1982.

Hardenability

The term "hardenability" is used freely throughout the remainder of this book; thus, it is important that the reader understand its precise meaning. One might think that hardenability means the ability to be hardened. Although true to a degree, the precise meaning of hardenability is much broader. In student classes on heat treating, two questions are often asked: (1) a specific steel, after heating above A_3 and quenching in water, shows a hardness of 65 HRC (near the limit for most steels as measured by Rockwell testing); does this establish it as a high-hardenability steel? and (2) after heating and quenching, a specific steel registers only 35 HRC; is it a low-hardenability steel? Invariably a student will answer yes to the first question and no to the second. The correct answer for both is "not necessarily." Information on section thickness in each instance is required before either question can be correctly answered. Therefore, hardenability does not necessarily mean the ability to be hardened to a certain Rockwell or Brinell value.

For example, just because a given steel is capable of being hardened to 65 HRC does not necessarily mean that it has high hardenability. Also, a steel that can be hardened to only 40 HRC may have very high hardenability. Hardenability refers to capacity of hardening (depth) rather than to maximum attainable hardness values.

As stated above, carbon controls the maximum attainable hardness for any conventional steel, but carbon has only a minor effect on hardenability. Thus, Fig. 11 also serves as an excellent example of a low-hardenability steel. Plain carbon steels are characterized by their low hardenability, with critical cooling rates lasting only for brief periods. Hardenability of all steels is directly related to critical cooling rates. The longer the time of critical cooling rate, the higher the hardenability for a given steel, almost regardless of carbon content (see TTT curve, Fig. 6, in Chapter 2).

Alloying Elements and Hardenability

The principal reason for using alloying elements in the standard grades of steel is to increase hardenability. The alloying elements used in conventional steels are confined to manganese, silicon, chromium, nickel, molybdenum, and vanadium. Because of the small amounts used, boron is not usually called an alloy. Steels that contain boron, whether they are carbon or alloy grades, are more often termed "boron-treated" steels. The use of cobalt, tungsten, zirconium, and titanium is generally confined to tool or other specialty steels. Tool steels and other highly alloyed steels are covered in other chapters of this book.

Manganese, silicon, chromium, nickel, molybdenum, and vanadium all have separate and unequal effects on hardenability. However, the individual effects of these alloying elements may be completely altered when two or more of these are used together. In periods of alloy shortages, extensive investigations were conducted, and it has been established that more hardenability can be attained with less total alloy content when two or more alloys are used together. This practice is clearly reflected in the standard steel compositions (see Chapter 6). This approach not only saves alloys that are often in scarce supply, but also results in more hardenability at lower cost.

Therefore, all of the alloy steels listed in Chapter 6 have significantly greater hardenability than the carbon steels. It must be further emphasized that the hardenability varies widely among the alloy grades, which is a principal reason for the existence of so many grades.

The quantitative effect of alloying elements has been determined for each of the elements commonly used. The relative effects of the various elements are expressed as multiplying factors based on percentage of the element, as shown in Fig. 12. When the perentage of each element is known (including the amount of carbon) for a specific steel, it is possible to calculate its hardenability in terms of ideal diameter, which is the diameter of a given steel specimen that will harden to 50% martensite at the center. Such calculations are somewhat complex and space-consuming; therefore, this aspect shall not be pursued further in this book. For further information on precise calculation of hardenability, see Ref 8.

Methods of Evaluating Hardenability

Many test methods have been devised for evaluating hardenability, although a majority of these test procedures are applicable for only a few (or sometimes only one) grades of steel. Other tests are arbitrary in nature and are not discussed in this book.

One hardenability test that has proved to be useful and reproducible is the hardness penetration diagram test. This is accomplished by plotting the cross-sectional hardness of a bar, or other section, on suitable graph paper. In this procedure, the specimen is hardened and sectioned. Hardness is then determined at various depths along the radius and plotted against distance from the surface, as in Fig. 13(a). Several hardness traverses are made along a number of radii, and the various values at each point are averaged before being plotted. After plotting, the mirror image is used to complete the symmetrical curve shown in Fig. 13(b). The pertinent point of this procedure is that a perfectly symmetrical curve results, based on an average of several readings. The hardness penetration diagram is adaptable to shallow, medium, or deep hardening steels.

A refinement of the hardness penetration diagram is designated as the S-A-C test. In this procedure, a 1-in.- (25-mm-) diam bar is quenched under standardized conditions, and the resultant hardness distribution is plotted as a symmetrical U-shaped curve. From this curve, the surface hardness is reported as S, the area under the curve as A (the units of area are "Rockwell

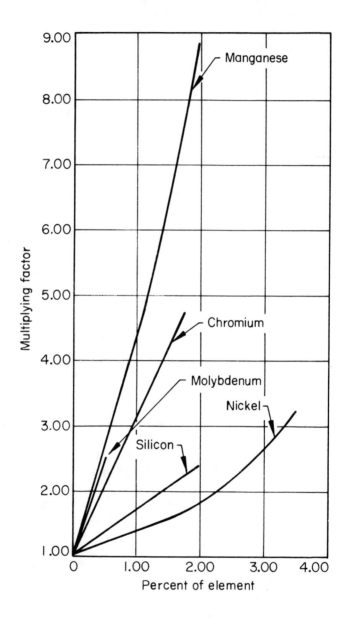

Fig. 12 Multiplying factors for five common alloying elements.
Source: MEI Course 10, Lesson 7, American Society for Metals, p
3, 1977.

inches"), and the hardness at the center of the piece as C. The test is commonly known as the Rockwell inch test. Because this test employs 1-in.- (25-mm-) diam bars, its usefulness is limited to steels that do not harden throughout such a section. If deeper hardening steels are to be tested by this procedure, use of a larger round is necessary.

End-Quench Testing. While the simple test as described above, and several other hardenability tests, are commonly used, the end-quench test is by far the most generally accepted and widely used method for evaluating the hardenability of carbon and alloy steels. The test is relatively simple to perform and can produce much useful information for the designer, as well as the fabricator.

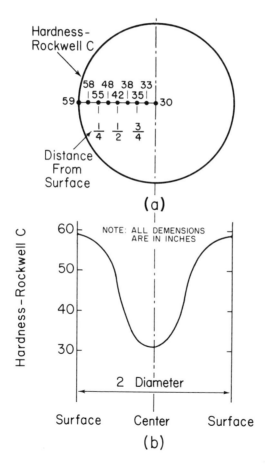

Fig. 13 Hardness penetration diagram. (a) Method of taking hardness traverse. (b) Plot of results of averaging several hardness traverses and mirror image. Source: MEI Course 10, Lesson 7, American Society for Metals, p 12, 1977.

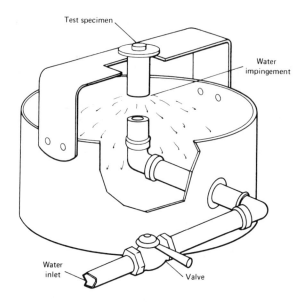

Fig. 14 Standard end-quench (Jominy) test specimen and method of quenching in quenching jig. Source: *Heat Treaters' Guide: Standard Practices and Procedures for Steel*, American Society for Metals, p 21, 1982.

Although variations are sometimes made to accommodate specific requirements, the test bars for the end-quench test are normally 1 in. (25 mm) in diameter by 4 in. (100 mm) long. A 1⅛-in. (28.5-mm) diameter collar is left on one end to hold it in a quenching jig, as illustrated in Fig. 14.

In this test, water flow is controlled by a suitable valve, so that the amount striking the end of the specimen (Fig. 14) is constant in volume and velocity. The water impinges on the end of the specimen only, then drains away. By this means, cooling rates vary from a very rapid rate on the quenched end to a very slow rate, essentially equal to cooling in still air, on the opposite end. This results in a wide range of hardnesses along the length of the bar.

After the test bar has been heated and quenched, two opposite and flat parallel surfaces are ground along the length of the bar to a depth of 0.015 in. (0.380 mm). Rockwell C hardness determinations are then made every 1/16 in. A specimen-holding indexing fixture is helpful for this operation for convenience as well as accuracy. Such fixtures are available as accessory attachments for conventional Rockwell testers. The next step is to record the readings and plot them on graph paper to develop a curve, as illustrated in Fig. 15. By comparing the curves resulting from end-quench tests of different grades of steel, their relative hardenability may be established. The steels having higher hardenability will be harder at a given distance from the quenched end of the specimen than steels having lower hardenability. Thus, the flatter the curve, the greater the hardenability. On the end-quench curves, hardness is not usually measured beyond approximately 2 in. (50 mm), because hardness measurements beyond this distance are seldom of any significance. At about this 2-in. (50-mm) distance from the quenched end, the effect of water on the quenched end deteriorates, and the effect of cooling from the surrounding air becomes significant. An absolutely flat curve demonstrates conditions of very high hardenability, which characterizes an air-hardening steel such as some of the highly alloyed tool steels.

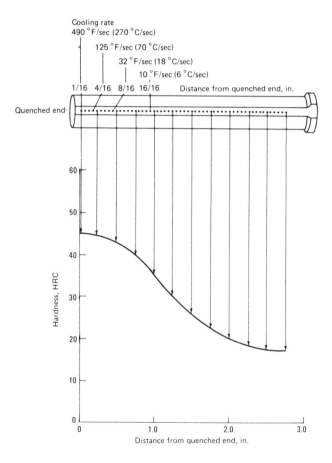

Fig. 15 Method of developing end-quench curve by plotting hardness versus distance from quenched end. Hardness plotted every 1/4 in. for clarity, although Rockwell C readings were taken in increments of 1/16 in., as shown at top of illustration. Source: *Heat Treaters' Guide: Standard Practices and Procedures for Steel*, American Society for Metals, p 21, 1982.

Variations in Hardenability. Because hardenability is a principal factor in steel selection and because hardenability varies over a broad range for the standard carbon and alloy steels, many grades are available. As a rule, hardenability of the standard carbon grades is very low, although there is still a great deal of variation in hardenability among the different grades. This variation depends to a great extent on the manganese content and sometimes to a smaller extent on the residual alloys that are sometimes present. A hardenability curve for a high-manganese grade of carbon steel, 1541, is shown in Fig. 16. This curve represents near maximum hardenability that can be obtained from any standard carbon grade.

In contrast to the curve shown in Fig. 16, typical hardenability curves for four different 0.50% carbon alloy steels are presented in Fig. 17. These data

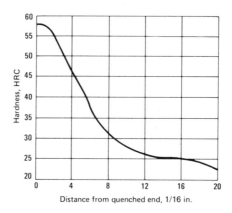

Fig. 16 End-quench hardenability curve for 1541 carbon steel. Source: *Heat Treaters' Guide: Standard Practices and Procedures for Steel*, American Society for Metals, p 22, 1982.

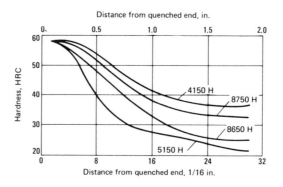

Fig. 17 Hardenability curves for several alloy steels. Source: *Heat Treaters' Guide: Standard Practices and Procedures for Steel*, American Society for Metals, p 22, 1982.

emphasize the fact that maximum attainable hardness is provided by the carbon content, while the differences in alloy content markedly affect hardenability.

H-Steels

Because of the normal variations within prescribed limits of composition, it would be unrealistic to expect that the hardenability of a given grade would always follow a precise curve, such as is shown in Fig. 16 and 17. Instead, the hardenability of any grade varies considerably, which results in a hardenability band such as the 4150H band in Fig. 18. This steel was normalized at 1600 °F (870 °C), then austenitized at 1550 °F (845 °C) before end quenching. The upper and lower curves that represent the boundaries of the hardenability band not only show the possible variation in hardness at the quenched end, caused by the allowable carbon range of 0.47 to 0.54%, but also the difference in hardenability as a result of the alloying elements being on the high or low side of the prescribed limits.

The need for hardenability data for steel users has long been recognized. Cooperative work by AISI and Society of Automotive Engineers (SAE) has been responsible for devising hardenability bands for a large number of carbon and alloy steels, principally the latter. Steels that are sold with guaranteed hardenability bands are known as the H-steels. The numerical parts of the designation are the same as for the other standard grades, but the

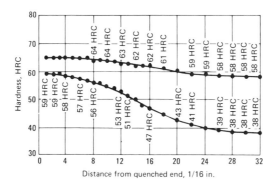

Fig. 18 Hardenability band for an alloy steel. 4150H: 0.47 to 0.54% C, 0.65 to 1.10% Mn. Normalized at 1600 °F (870 °C). Annealed at 1550 °F (845 °C). Source: *Heat Treaters' Guide: Standard Practices and Procedures for Steel*, American Society for Metals, p 22, 1982.

suffix letter H, such as 4140H, identifies it as a steel that meets prescribed hardenability limits. Not all of the steels listed by AISI are available as H-grades.

In order to give steel producers the latitude necessary in manufacturing for common hardenability limits, the chemical compositions of normal grades have been modified to form the H-steels. These modifications permit adjustments in manufacturing ranges of chemical composition. These adjustments correct melting practice for individual plants which might otherwise influence the hardenability bands. However, the modifications are not great enough to influence the general characteristics of the original compositions of the steels. Figure 18 is presented merely as an example of a hardenability band.

Chapter 4

Furnaces and Related Equipment
for
Heat Treating

Heat treating furnaces and related equipment include the heating devices (furnaces), fixtures and/or holding devices, quenching systems, and atmosphere and temperature control systems—all of which are required for the majority of heat treating operations. Specialized equipment for induction and flame heat treating are covered in Chapter 11.

Types of Heat Treating Furnaces

Sizes and designs of heat treating furnaces vary over such a wide range that any precise classification is virtually impossible. In size, furnaces vary from a small model that sits on a bench and has a work space capacity for only a few ounces (often used for heat treating instrument parts) to a large car-bottom furnace that is capable of handling hundreds of tons in a single heat (Fig. 1).

Regardless of size, furnaces may be directly fired with fuel, where the work is exposed to combustion gases, or indirectly fired where the work is separated from combustion gases. Furnaces may also be heated by electrical resistance.

Ovens and Furnaces

The Industrial Heating Equipment Association (IHEA) classifies heating devices as ovens and furnaces. This separation is made on the basis of operating temperature—up to about 1000 °F (540 °C) is an oven, and any unit that operates at temperatures exceeding 1000 °F (540 °C) is called a furnace. This separation, based on operating temperature, is related directly to heating mode.

Fig. 1 Car-bottom furnace used for stress relieving large weldments. Source: MEI Course 6, Lesson 6, American Society for Metals, p 13, 1977.

Modes of Heating (Ref 9)

The three basic modes of heat transmission are conduction, convection, and radiation. In industrial heat treating, these modes may be used singly, or in combination.

Conduction of heat in a solid such as a metal workpiece is the transfer of heat from one part of the solid to another, under the influence of a temperature gradient and without appreciable displacement of the particles. For instance, if the temperature of the surface of a part is elevated, the heat flow to the center is by a molecular mechanism. Conduction involves the transfer of kinetic energy from one molecule to another in a chain reaction. Heat flow continues until equilibrium occurs. The time involved depends on the conductivity of the given metal, but in general, the speed of conduction within metals is relatively fast.

In most heat treating processes, while conduction plays only a minor role in the total heat transfer from the source to the workpiece, it is the sole mode of transferring heat from the surface to the center of a workpiece. An exception to the minor role conduction normally plays is the immersed electrode salt bath or a fluidized bed. While all three modes of heating are utilized in a molten salt, molten metal bath, or a fluidized bed, conduction plays an important role because the hot medium is in direct contact with the work-metal surfaces.

Convection involves the transfer of heat by mixing one parcel of fluid (fluid refers to either liquid or gas) with another. The motion of the fluid may be entirely the result of density differences resulting from temperature difference, as in natural convection, or it may be produced by mechanical means, as in power convection. Fans commonly are used to increase the overall heat transfer coefficient of the system. A hot-air home heating system is an excellent example of heating by convection.

Tempering of steel is a common application of convection heating in heat treating. A typical installation is shown in Fig. 2. This furnace is heated by recirculating, forced air convection. The electric heating elements are located apart from the work chamber. The air in the furnace is forced through the heating chamber at high velocity and then through the work chamber. With this system, accurate control of temperature is easily achieved. Also, for any heat treating application (tempering or other), this method of heating is highly efficient up to about 900 °F (480 °C). It is used for processing at somewhat higher temperatures although efficiency of convection heating decreases as the temperature is increased beyond approximately 900 °F (480 °C). Most ovens, as defined by IHEA are heated by the convection mode.

Radiation. A body emits radiant energy in all directions by means of electromagnetic waves; the wave length ranging from 4 to 7 μm. When this energy strikes another body, some of the energy is absorbed, raising the level

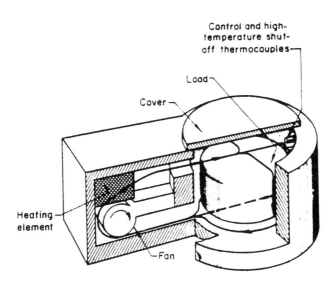

Fig. 2 Low-temperature heat processing furnace heated by convection. Source: MEI Course 20, Lesson 4, American Society for Metals, p 3, 1977.

of molecular activity and producing heat. Some of the energy is reflected. The amount absorbed depends on the emissivity of the surface of the receiver. The sender gives up heat or energy. On this basis, if two pieces of metal, one hot and one cold, are placed in a completely insulated enclosure, the hot piece cools and the cold one is heated. The exchange of energy takes place until both objects come to equilibrium or to the same temperature level. Even after equilibrium of temperature is established, the process continues with each piece radiating and absorbing energy from each other.

Transfer of heat by radiation, therefore, relates directly to emissivity, which is the ratio of loss of heat per unit area of a surface at a given temperature to the rate of heat loss per unit area of a black body at the same temperature in the same surroundings. The practical meaning is that when a workpiece is placed in a furnace and exposed to radiant heat, its rate of heating depends on its surface. A highly reflective object (polished stainless steel, for example) absorbs heat at a lower rate compared with a dark workpiece.

As a rule, in practical heat treating, no attempt is made to alter the workpiece surfaces to make them more receptive to radiant energy, although workpieces have been intentionally blackened to improve heating efficiency.

Most heat treating furnaces that operate at temperatures higher than approximately 1100 °F (595 °C) are heated primarily by radiation. This is generally true regardless of their size, or whether they are heated by electrical resistance elements, directly by means of radiation from burners and the furnace walls, or indirectly by tube-contained burners (radiant tubes). A typical radiation-heated heat treating furnace is the simple box-type batch furnace shown in Fig. 3. Electrical resistance elements can be seen on the side walls of the furnace. However, a convection assist is employed in many larger heat treating furnaces; that is, circulating fans are used to increase heating efficiency and temperature uniformity.

Classification of Furnaces by Heat Transfer Medium

One means of classifying heat treating furnaces is by the type of heat transfer medium employed; this classification method is valid regardless of size and most common variables of furnace components.

Until recently, only two types of heat transfer media—gaseous (air or vacuum) and liquid (molten metal or molten salt bath)—were commonly used. Fluidized bed furnaces now have proved to be a useful tool in heat treating, adding a new method of heat transfer (solid). Those classifications, for the most part, are used in the remainder of this chapter.

Many types and designs of heat treating furnaces, regardless of the mode of heat transfer or the medium employed for heat transfer, are available as standard models. When an existing predesigned and/or prebuilt model is suitable for specific customer requirements, cost is naturally lower, because engineering has been completed.

Fig. 3 Small batch-type furnace. Note steel frame, insulating brick, electrical resistors and roll-top table in front of furnace for handling work. Source: MEI Course 6, Lesson 6, American Society for Metals, p 11, 1977.

In many instances, standard models require minor factory modifications to meet specific customer requirements. This increases cost, but seldom equals the cost of designing and building a custom furnace. However, hundreds of heat treating furnaces exist that have been designed and constructed for a specific application; notably the larger furnaces (Fig. 1).

Batch-Type Versus Continuous-Type Furnaces

In some plants, heat treating furnaces are classified as batch or continuous types. A batch-type furnace refers to one that is loaded with a charge and then closed for the pre-established heating cycle. After completion of the heating cycle, the work load may be cooled in the furnace at a planned cooling rate (such as for annealing), removed, and cooled in still air (as in normalizing), or quickly cooled (quenching) as by immersion in oil or water.

Figures 1, 2, 3 and 4 illustrate specific types of batch-type furnaces. The large furnace shown in Fig. 1 is a large car-bottom type.

Car-bottom furnaces incorporate a rail car for the hearth of the furnace. This hearth must be well insulated to keep the heat from reaching the wheel roller surfaces. The furnace usually is built at floor level and frequently is equipped with a lift door. The work to be heated is placed on the hearth, the car moved into the furnace on rails, and the door closed. Troughs filled with sand usually provide a seal between the lower edges of the furnace and the car bottom. Another variation of this design is the elevator car-bottom furnace, where the furnace shell may be lifted while the car bottom is being positioned. The furnace shell is then lowered and rests on the car hearth during the heating period. The problem of proper sealing is greatly simplified with this construction. The primary disadvantage of the car-bottom furnace is that the furnace cannot be heated with the car out of position, because the latter forms the insulated floor of the furnace. Such furnaces are most commonly used for very large workpieces that require stress relieving, annealing, or normalizing.

Another batch-type furnace is shown in Fig. 2. This unit is heated by convection and frequently is used for tempering, stress relieving, or process (subcritical) annealing. As indicated in Fig. 2, this type of furnace is a top-loader. As a rule, furnaces of this type are relatively small, but there is no specific size limitation.

Figure 3 illustrates a batch-type furnace which is one of the real "work horses" of the heat treating industry. Regardless of the type of energy employed for heating, the heating mode is principally radiation. Such furnaces are generally available in a broad range of sizes and can be heated with gas, oil, or electricity. As shown in Fig. 3, it is heated by electrical heating elements (metallic), which allows efficient operation up to 1800 °F (980 °C) or slightly higher. However, by use of silicon-carbide elements, the operating temperature can be extended to 2300 °F (1260 °C) or slightly higher, thus permitting use for heating highly alloyed tool steels (see Chapter 8). Therefore, such a furnace is quite versatile and can be used for almost any heat treating operation. However, it generally is not recommended for heating at temperatures lower than approximately 1200 °F (650 °C), unless equipped with one or more circulating fans. This type of furnace can be operated with a "natural" atmosphere or with any one of several prepared atmospheres as required.

Another example of a popular, standard-design batch-type furnace is shown in Fig. 4, a pit-type vertical loader that can be adapted to almost any heat treating operation. The work area is surrounded by gas-heated radiant tubes (indirectly fired). A convection assist is supplied by a bottom fan. Furnaces of this type also can be heated by electrical resistance and are amenable to use with a variety of prepared atmospheres.

Bell-type furnaces (not illustrated) are also widely used as batch-type units. In bell-type furnaces, the round hearth or base that supports the load is stationary at floor level, while the furnace can be lifted off and transferred from one base to another by an overhead crane. After the work is placed on the base, the furnace is lowered and properly positioned on the base by guideposts. Sealing is effected by sand or oil in a circular trough around the outside of the base. Often an inner metal retort or muffle with a skirt at the bottom to seal in a protective atmosphere is first used to cover the work before the furnace is lowered into place.

This type of furnace is widely used with an inner muffle and protective atmosphere for annealing material in coil form, such as steel sheet or wire, and nonferrous products.

Continuous furnaces may be any of many designs including their conveying mechanisms, but basically they portray an "in-one-end" and "out-the-other-end" type of unit. A continuous furnace is generally intended for continuous high production of similar parts. The capabilities that can be designed into a continuous furnace are virtually limitless in terms of varied heat treating cycles.

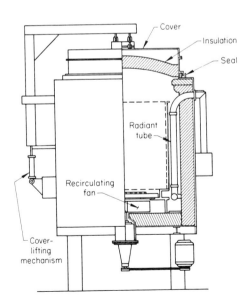

Fig. 4 A pit batch furnace. Dashed lines outline location of workload. Source: *Carburizing and Carbonitriding*, American Society for Metals, p 55, 1977.

Fig. 5 Roller hearth furnace showing charging end. Rolls leading into the furnace for transporting work are similar to those throughout the length of the furnace. Source: MEI Course 6, Lesson 6, p 15, 1977.

Roller-Hearth Continuous Furnaces. The charging end of a large continuous furnace is illustrated in Fig. 5. As shown in Fig. 5, the work is conveyed through the furnace by means of rollers. The ends of the rolls project through the walls of the furnace to external air- or water-cooled bearings. Usually, the rolls are power driven by a common source through a chain and sprocket mechanism. Frequently, the driven rolls extend to some distance beyond the furnace at both the loading and discharging end.

The work is placed on the revolving rollers at the charge end and is carried through and out of the furnace by the friction between the work surface and the revolving roll surfaces. This obviously works only if the charge material is sufficiently long relative to the roll spacings. Work of smaller size can be stacked on trays, loaded into baskets, or hung on special fixtures carried by the rollers. Figure 5 shows the entrance end of a roller hearth furnace fired with radiant-tube burners. The entrance vestibule is equipped with a labyrinth of vertical asbestos curtains to help confine the internal protective atmosphere. A similar vestibule is provided at the discharge end when work is cooled inside the furnace.

Belt-Type Continuous Furnaces. In this construction, a continuous conveyor belt acts as a moving hearth to carry the work through the furnace. Several belt designs are used, depending on the size and weight of the work to be handled, how the hot belt is supported inside the furnace, and the operating temperature. Belts may be constructed of woven wire mesh, flat cast alloy links, or a more open slat design. They may have a relatively smooth even surface, or may have surfaces that are recessed, grooved, or channeled to control spacing between individual parts of the load. The conveyor usually is driven by large diameter drums at each end of the furnace. The belt may return either outside or inside the furnace, depending on the function the furnace is performing. This type of furnace is suited to continuous annealing, tempering, sintering, and hardening operations. If desired, the heated workpieces may be dumped directly into an enclosed quenching tank without losing atmosphere protection.

Pusher-type continuous furnaces are usually designed to carry higher unit loads than belt-type furnaces. The work is placed on the hearth of the furnace and pushed ahead periodically by a mechanical ram operating at the charging end. Work may be loaded onto sturdy cast alloy trays, baskets, or other fixtures that ride on skid rails, tracks, or rollers built into the floor of the furnace. The pusher can be actuated by an air or hydraulic cylinder, or a rack and pinion mechanism driven by an electric motor. The action of the ram and opening and closing of interior and/or exterior furnace doors must be properly coordinated by an interlocking safety system.

Rotary-hearth continuous furnaces are exception to the "straight through" type of continuous furnace, as illustrated in Fig. 6. In this type of furnace, the hearth is a flat ring, similar to the revolving floor of a merry-go-round. The furnace in Fig. 6 is heated with gas-fired radiant tubes, but such a furnace can be heated electrically. As also indicated in Fig. 6, the work is charged into and removed through a single opening. The length of the heat cycle is governed by establishing the speed of hearth rotation. While there are no fixed limitations on applications of rotary hearth furnaces, they usually are used for heating large workpieces.

Liquid Bath Furnaces

Heating by immersing the workpieces in a liquid represents an entirely different concept compared to gaseous atmosphere furnaces. Heating parts in molten metal (usually molten lead) is an age-old practice. A pot-type furnace containing molten lead provides an effective means of heating steel parts, but there are certain disadvantages:

- Molten lead is heavy; consequently, steel parts float if not anchored.
- Lead adheres to steel parts, which impairs the quenching action and poses a cleaning problem.

A few lead-pot furnaces are still being used, but for the most part, they have been replaced by molten salt bath furnaces.

Molten salt baths offer several distinct advantages: (1) salts are available for operation in the temperature range of 350 °F (175 °C) to 2300 °F (1260 °C); (2) parts do not scale or otherwise result in deteriorated surfaces, because

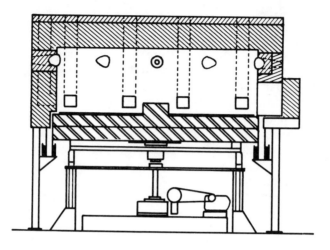

Fig. 6 Rotary hearth forging or heat-treating furnace. Source: MEI Course 6, Lesson 6, p 6, 1977.

they are fully protected while they are in the molten bath; (3) a thin film of salt remains on the work during transfer from heat to quench so that clean hardening is facilitated; and (4) a wide variety of salts are available, including salts that change the surface chemistry of the steel (see Chapter 10).

The principal disadvantage of heating parts in molten salt is the necessary cleaning after heat treating, which can be difficult, especially for parts of complex design.

Types of Salt Baths. The simplest form of molten salt bath involves heating a metal pot filled with a low-melting-temperature salt with electric immersion heaters. This type of salt bath, however, is applicable only to temperatures from approximately 350 to 650 °F (175 to 345 °C).

As temperature requirements increase, more sophisticated equipment is required. A common type of fuel-fired pot furnace that can be used for either molten metal (usually lead) or molten salt is shown in Fig. 7(a). This type of furnace can be used for heating metals up to about 1650 °F (900 °C). Higher temperatures can be achieved, but deterioration of the pot and other components becomes excessive.

The furnace shown in Fig. 7(b) is similar to Fig. 7(a), except it is heated by electrical resistance. Because the pot in this type of furnace is constantly surrounded with a strongly oxidizing atmosphere, pot life is shortened when used in the higher temperature ranges.

Both of the above-mentioned types of furnaces are extremely versatile, but they are best suited to limited production of a variety of small parts. To attain

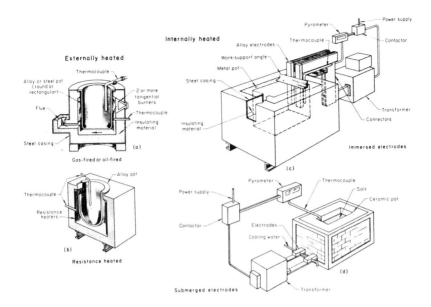

Fig. 7 Principal types of salt bath furnaces. Types (a) and (b) also can be used for lead baths. Source: MEI Course 6, Lesson 6, p 17, 1977.

acceptable pot life, the pots must be made from an expensive nickel-chromium alloy material.

Most liquid bath heat treating requiring the temperature range of about 1400 to 2300 °F (760 to 1260 °C) is done in furnaces such as those illustrated in Fig. 7(c) and (d), immersed and submerged electrode types, respectively. In both types, heat is generated by resistance to current flow through the molten salt from one electrode to the other. This creates a stirring action at the electrodes, thus providing uniform temperatures within the bath.

The two types (immersed and submerged electrode) generally are competitive with each other; each having certain advantages and disadvantages. Either type lends itself to batch or continuous operation.

A principal disadvantage of the immersed electrode type (over-the-top electrodes shown in Fig. 7c) is that the electrodes deteriorate just above the salt line where the heat is intense. The submerged electrode type overcomes this disadvantage; however, the submerged type furnace requires a molded-in ceramic type of pot, and ceramic pots are not compatible with all salt compositions.

Salt bath furnaces heat by radiation and conduction. In the immersed and submerged electrode types, convection heating is also added because of the stirring action. Therefore, heating rate in any salt bath is much greater than in the gaseous atmosphere furnaces discussed earlier in this chapter.

Molten salt baths do, therefore, offer an efficient means of heating metals—principally steels—although some salt compositions are compatible with nonferrous metals and alloys. The degree of economy in heating is particularly realized with the immersed and submerged electrode types, if they can be properly applied. Both of these furnace types are best adapted to continuous production because of the difficulty in restarting them when salts are allowed to solidify. Thus, such furnaces are not well suited to intermittent production. For weekends or other required downtime, they are idled at a temperature just above the solidification temperature of the salt.

Fluidized Bed Furnaces (Ref 10 and 11)

The newest approach to heating metals is by means of the fluidized bed. Fluidized bed heating is carried out in a bed of mobile inert particles, usually aluminum oxide. These particles are suspended by the combustion of a fuel/air mixture flowing upward through the bed. Components are immersed into the fluidized bed as if it were a liquid and are heated by the hot fluid bed. Heat transfer rates in a fluidized bed are up to ten times that which can be achieved in conventional direct-fired furnaces. The combination of high heat transfer, excellent heat capacity, and uniformity of behavior over a wide temperature range provides a constant temperature bath for many applications.

Heat treatment in fluidized beds was originally patented in 1950 by British Aluminum. Until recently, however, it was only possible to heat fluidized beds electrically, making their use for metal processing at temperatures above 1300 °F (705 °C) difficult and inefficient. The advent of fuel-fired fluidized beds using a gas/air mixture as both the heating and fluidization medium has altered this condition. As a result, fluidized bed furnaces are available to perform standard heat treating operations.

A fluidized bed is held in a metallic or refractory container and is started by initially lighting the combustion mixture at the top. The flame-front gradually moves down the depth of the bed until it stabilizes above the ceramic plate. The combustion occurs spontaneously within about 1 in. of the plate surface. The ceramic distribution plate ensures uniform properties within the bed.

There are two types of fluidized beds available, internally fired for high-temperature applications (1400 to 2220 °F or 760 to 1215 °C), and externally fired for low-temperature applications (1400 °F or 760 °C and below).

In the internally fired bed (Fig. 8a), the fuel and air are mixed in near stoichiometric proportions and passed through a porous ceramic plate above which the particles are fluidized in the gas stream. The gas stream imparts its thermal energy to the bed particles which in turn impart thermal energy to the object being processed.

For processing temperatures below 1400 °F (760 °C), the externally fired bed is used. Such a bed is illustrated in Fig. 8(b). In this system, an excess air

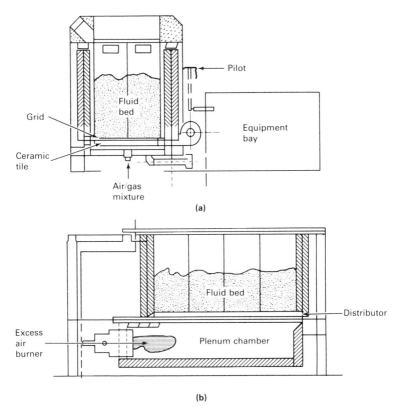

Fig. 8 Sectional views of fluidized bed furnaces. (a) Internally fired. (b) Externally fired. Source: Boyer, H.E., ASM Symposium, 26 March 1981, Vancouver, BC.

burner fires into a plenum chamber, above which the fluidized bed is supported by a porous metallic plate. The bed is fluidized by the products of combustion from the plenum chamber.

The temperature of the bed is automatically controlled by a proportioning controller linked to a motorized valve that meters the appropriate amount of gas/air mixture to the distribution tile. The control is arranged so that its range is well above the minimum fluidization velocity of the particles. The fluidization gas is also the furnace atmosphere in which the metal is treated. By regulating the air/gas ratio to the distributor tile, the atmosphere may be inert, oxidizing, or reducing.

Applications for fluidized beds are numerous. The most obvious in heat treating are neutral hardening, where they may be used in place of neutral salt baths because the rate of heat transfer is even faster than in any salt bath. Additionally, particles from the fluidized bed do not adhere to workpieces so that there is no cleaning problem. There is no dragout of particles from a fluidized bed, compared to the constant dragout (and need for replenishment) of salt from a molten salt bath furnace. Therefore, this can be a significant cost factor in favor of the fluidized bed. Actually, fluidized beds can be adapted to virtually all heat treating operations, which include hardening of steel (even high-speed tool steels), normalizing, annealing, stress relieving, and tempering. Fluidized beds are also well suited to heat treating of nonferrous metals.

Vacuum Furnaces

Heating of metal parts in a vacuum furnace represents a relatively new development in the metallurgical field. It consists of carrying out various thermal operations in a heated chamber evacuated to a vacuum pressure suitable to the particle material and process desired. Although originally developed for the processing of electron-tube and space-age materials, it has been found to be extremely useful in many less exotic metallurgical areas as vacuum technology has progressed. Vacuum heat treating can be used to:

- Prevent reactions at the surface of the work, such as oxidation or decarburization, thus retaining a clean surface intact
- Remove surface contaminants such as oxide films and residual traces of lubricants resulting from fabricating operations. The latter often are severe contaminants to the furnace
- Add a substance to the surface layers of the work, such as by carburization
- Remove dissolved contaminating substances from metals, using the degassing effect of a vacuum, such as hydrogen or oxygen from titanium
- Join metals by brazing or diffusion bonding

Operations performed in vacuum furnaces include hardening, tempering, annealing, stress relieving, and sintering. In vacuum furnaces, a complete vacuum (absolutely none of the original air) is virtually impossible to attain. One standard atmosphere at sea level equals 760 mm of mercury. The degree of vacuum used for most heat treating operations is about 1/760 of an atmosphere. Under these conditions, the amount of original air remaining in the work chamber is approximately 0.1%. This degree of vacuum can normally be obtained by "pumping down" with a mechanical pump. There are, however, some heat treating operations involving highly alloyed materials that require a "harder" vacuum; that is, less than 0.1% of the original air. Under these conditions, mechanical pumping is followed by use of the highly sophisticated oil-diffusion pump.

One reason for the relatively slow acceptance of the vacuum furnace for heat treating was that originally very rugged construction was required for building a hot-wall furnace (no water cooling of the exterior). Construction materials, including the steel shell, lose their strength at elevated temperature;

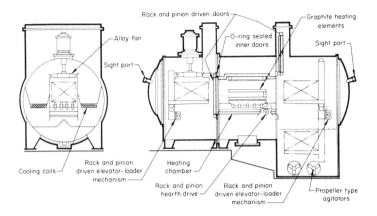

Fig. 9 Three-chamber cold-wall vacuum oil quench furnace. Source: MEI Course 6, Lesson 7, p 17, 1977.

consequently, wall thicknesses used for conventional heat treating furnaces were susceptible to implosion under vacuum.

This problem was solved by the advent of the cold-wall (water-cooled walls) furnace. Side and end sectional views of a cold-wall model are presented in Fig. 9, which is a three-chamber furnace that includes a loading vestibule (left of the side view), heating chamber (center of side view), and an elevator quenching arrangement at the extreme right of the side view. Vacuum furnaces are heated by electric resistors, frequently the graphite type as indicated in Fig. 9.

Vacuum furnaces offer a number of advantages, which include certain economies in operation; actually, a vacuum atmosphere is about the cheapest prepared atmosphere. High initial cost and requirements for a measure of skill in operation are presently the principal disadvantages.

Furnace Parts and Fixtures

While the interior components of heat treating furnaces are, insofar as possible, constructed of refractory materials, there are many components, especially in highly sophisticated systems, that are necessarily made from metal. Compositions of the metal parts may vary from simple low-carbon steel (0.15 to 0.20% C) up to and including highly alloyed, expensive metals that contain up to 60% Ni and 25% Cr. Carbon steels not only scale excessively at temperatures higher than about 800 °F (425 °C), but they lose their strength, so that they are essentially worthless for furnace parts that operate at elevated temperatures. Type 310 stainless steel (25% Cr and 12% Ni) is used for furnace parts where temperatures do not exceed about 1200 °F

(650 °C), but higher temperatures demand the more highly alloyed materials for reasonable life. An alloy that can be cast easily, can be obtained in wrought form, and can be welded, contains approximately 35% Ni and 18% Cr. Such an alloy provides reasonable life at temperatures of 1700 °F (925 °C) or even higher. The most efficient and economical material for furnace parts depends on a number of factors; for more information, see Ref 10.

Fixtures and baskets used for heat treating are a necessary evil; frequently, the baskets or fixtures required for holding the workpieces weigh more or are larger than the work loads they contain. Under these conditions, more energy is used to heat and cool the fixtures or baskets than is used to process the workpieces. Therefore, any heat treater prefers to avoid the use of fixtures, baskets, or trays whenever possible. In many instances, the workpieces can be placed so that fixtures are not required (note in Fig. 1). In many other instances, fixturing is absolutely essential to minimize distortion of the workpieces and to allow uniform cooling. Automotive parts such as gears, pinions, axles, and shafts are typical parts that require fixturing or arrangement in baskets for heat treating. The types of fixtures, trays, and baskets are essentially endless; most are designed for specific applications.

A typical multipurpose tray is shown in Fig. 10. This tray was fabricated by welding wrought components of a 35% Ni-18% Cr alloy. It can accommodate arrangement of a wide variety of parts and can be adapted to either batch-type

Fig. 10 Bar frame type basket. Source: *Metals Handbook*, Vol 4, 9th edition, American Society for Metals, p 330, 1980.

Fig. 11 Large pit furnace basket. Source: *Metals Handbook*, Vol 4, 9th edition, American Society for Metals, p 330, 1980.

or continuous furnaces. For processing salt baths or fluidized beds, such trays are designed to be higher and narrower, but design principles are similar.

For heat treating in pit-type furnaces (Fig. 4), a round basket or tray is most efficient, as shown in Fig. 11. This basket is made by casting from a heat-resistant alloy; it has seven trunions that are engaged by hooks from a gib crane to load and unload vertically from pit-type furnaces.

Design Criteria. In planning heat treating operations, the design and use of trays, fixtures, and baskets should attempt to:

- Eliminate their use whenever possible; every unit of weight consumed by fixtures reduces payload by an equal amount.
- When fixtures or baskets are necessary, the design of these fixtures should reduce weight to the minimum that will provide the required strength to handle the charge.
- In quenching applications, fixtures should be designed with openings to allow circulation of the quenching medium (see Fig. 10 and 11).

Suppliers of alloy fixtures—cast, wrought, and welded structures—have an array of standard designs that can be used as is or modified to suit specific heat treating applications.

Quenching Mediums and Systems (Ref 13)

Quenching is the rapid cooling of a metal or an alloy from a suitable elevated temperature. This usually is accomplished by immersion in water, oil, or a polymer, although forced or still air is sometimes used and is considered quenching.

There are two major applications of quenching. One is cooling of quench-hardenable steels to develop an acceptable as-quenched microstructure and mechanical properties that meet minimum specifications after the parts are tempered. The second is the rapid cooling of certain iron-base alloys and nonferrous metals from elevated temperatures to retain a uniform solid solution in the metal. Such treatments are used to obtain a fully annealed condition to permit forming or to permit precipitation hardening by a subsequent aging process. The major portion of this discussion covers quench hardening of steels.

Metallurgical Aspects. Most steels are quenched to control the transformation of austenite and to subsequently form the desired microconstituents. Microstructures that may be obtained are indicated on the combination TTT (time, temperature, transformation) diagram (see Chapter 2).

Martensite is the as-quenched microstructure that is usually desired, as indicated by curve A in Fig. 12. Along cooling curves B, C, and D, some transformation to bainite and ferrite occurs, with a corresponding decrease in the amount of martensite formed and the hardness developed. Formation of these mixed structures commonly is referred to as "slack quenching." See Chapter 2 for an in-depth discussion of TTT curves.

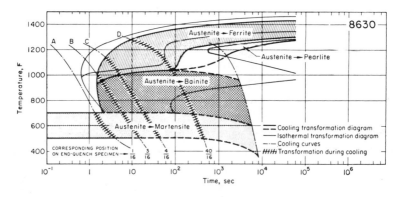

Fig. 12 Transformation diagrams and cooling curves for AISI 8630 steel, indicating the transformation of austenite to other constituents as a function of cooling rate. Source: MEI Course 6, Lesson 15, American Society for Metals, p 5, 1978.

Cooling Rates. When carbon steel is quenched from the austenitizing temperature, a cooling rate equal to or greater than about 100 °F/s (55 °C/s) (measured at 1300 °F, or 704 °C) is necessary for avoiding the nose of the TTT curve. The actual cooling rate required depends on the hardenability of the steel. The entire cross section of the workpiece must cool at this rate to obtain the maximum amount of martensite. Under ideal conditions, water provides a cooling rate of about 500 °F/s (278 °C/s) at the surface of steel cylinders 1/2 in. diameter by 4 in. long. This rate decreases rapidly below the surface. Thus for carbon steel, only light sections with a high ratio of surface area to volume can be fully hardened throughout the cross section.

Quenching Mediums

Many different mediums have been used for quenching. Most of them are included in the list which follows, and some of these are used only to a very limited extent:

- Water
- Brine solutions (aqueous)
- Caustic solutions
- Polymer solutions (synthetics)
- Oils
- Molten salts
- Molten metals
- Gases, including still or moving
- Fog quenching
- Dry dies, commonly water cooled

In cooling power, nonagitated water is arbitrarily rated as (1.0); other mediums are rated by comparison to this value. The relative cooling powers of the most common quenching mediums (water, oil, and brine or caustic solutions) are presented in Table 1, which also shows the effect of agitation on cooling power. With agitation, the cooling power of any medium is greatly increased. Cooling powers of synthetic quenching mediums (polymers) are not shown in Table 1 because they can vary from the relatively low power of oil to the high quenching power of brine. Synthetic materials are available as proprietary materials for mixing with water. Synthetic materials are somewhat difficult to use because quench bath composition is easily degenerated through dragout (the materials adhere to the quenched workpieces).

Table 1 shows that brine or caustic (usually 5 to 10% in water) is an extremely severe quenching medium. When strongly agitated, it has a quenching power of 5 and is, therefore, often preferred over water. A

Table 1 Effect of agitation upon the effectiveness of quenching
Source: MEI Course 6, Lesson 15, American Society for Metals, p 8, 1978.

Circulation or agitation	H-value or quenching power		
	Oil	Water	Caustic soda or brine
None	0.25-0.30	0.9-1.0	2
Mild	0.30-0.35	1.0-1.1	2-2.2
Moderate	0.35-0.40	1.2-1.3	...
Good	0.4-0.5	1.4-1.5	...
Strong	0.5-0.8	1.6-2.0	...
Violent	0.8-1.1	4	5

disadvantage of brine is its corrosivity to the equipment. Caustic is not usually corrosive, but can endanger personnel if not handled properly.

Water is inexpensive and is a very effective quenching medium. However, it is susceptible to formation of steam pockets which may result in soft spots on the quenched workpieces. This is particularly likely if the temperature of the quenching system is not closely controlled. In a water quenching system, the water should never be allowed to exceed 90 °F (32 °C).

Oil generally has a quenching power (depending on the type of oil) of approximately 0.25 to 0.40% of that of plain water, assuming similar agitation.

Oil is less temperature sensitive than water. Also, the most effective quenching temperature for oils begins at about 90 °F (32 °C), whereas this is the upper effective limit for water. This is due to the fact that oils are viscous and therefore have poor mobility at lower temperatures.

Other quenching mediums listed above generally are restricted to use for specialized applications. Air cooling (quenching) often is used for very high hardenability steels.

Quenching Systems

Equipment requirements for quenching may vary over a wide range. For instance, in a small shop where only a few small parts made from carbon steel are heat treated each day, the equipment might be as simple as a steel barrel containing water or brine. No heating or cooling facilities are necessary, and the agitation is accomplished by the operator moving the workpiece about in the medium during cooling. Such a simple system usually is inadequate.

At the other extreme, for continuous operation, the quenching system may contain all or most of the following components regardless of the quenching medium used:

- Tanks or quenching machines
- Agitation equipment
- Fixtures for quenching
- Cooling systems
- Storage or supply tanks
- Heaters
- Pumps

A simple, but effective, tank system for water quenching is shown in Fig. 13. In this system, a supply of fresh water is continuously fed into the bath and allowed to overflow to the drain.

The system is intended for use where water is plentiful and inexpensive. Where water is scarce, however, this system may be connected to a central or closed water-cooling system that would condition the water for reuse.

The ultimate in a sophisticated oil quenching system is illustrated in Fig. 14. All of the components listed above are included in the system design. Such installations are common where production rates are high and demands for quenching are essentially continuous.

Furnace Atmospheres

In many heat treating operations, austenitizing must be performed under conditions that provide some form of surface protection for the workpieces. If this is not done, workpieces may be severely oxidized (coated with rust) and

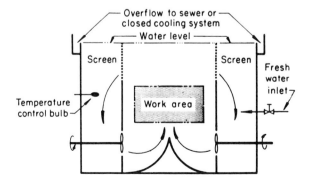

Fig. 13 Temperature-controlled overflow tank for water quenching. Source: MEI Course 6, Lesson 15, American Society for Metals, p 25, 1978.

decarburized, or both. As steel is heated, the surfaces become more active chemically as the temperature increases. Severe oxidation of carbon steels commonly begins at about 800 °F (425 °C). At temperatures exceeding approximately 1200 °F (650 °C), the rate of oxidation becomes exponential as temperature increases. Under high-temperature conditions, decarburization (a heat treater's curse) also occurs. The carbon in the steel reacts with the atmosphere so that the amount of carbon is substantially reduced (usually a highly undesirable condition). Generally, furnace atmospheres serve one of two requirements; they provide heat treated workpieces whose surfaces are clean and essentially unchanged from their preheated condition (neutral heating), and they serve to attain a controlled condition of surface change, as in certain case hardening operations. When workpieces are heated in salt baths or fluidized beds, atmospheres are automatically provided because workpieces are immersed. Controlled changes in the surface chemistry of the workpieces may also be accomplished in molten salts or fluidized beds.

Principal types of gaseous atmospheres that may be used in atmosphere furnaces are listed below in the general order of increasing cost:

- Natural (conventional air)
- Atmosphere derived from productions of combustion in a direct fuel-fired furnace
- Exothermic (generated)
- Endothermic (generated)
- Nitrogen base
- Vacuum
- Dissociated ammonia
- Dry hydrogen (dried bottled gas)
- Argon (from bottles)

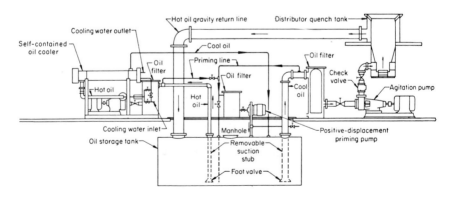

Fig. 14 Typical oil-cooling system employing an underground storage tank. Source: MEI Course 6, Lesson 15, American Society for Metals, p 29, 1978.

Discussion of furnace atmospheres can be extensive and complex. Therefore, at the risk of some oversimplification, an attempt is made below to briefly describe the types of atmospheres available, emphasizing applications and limitations. For more detailed information on furnace atmospheres, see Ref 10 and 12.

Natural Atmospheres (Air). A natural atmosphere is the air we breathe, which is essentially composed of 79% nitrogen and 20% oxygen. Such atmospheres exist in any heat treating furnace where the work chamber does not contain products of combustion, or where specially prepared atmospheres are not admitted (see Fig. 3). In many heat treating applications, this type of atmosphere is acceptable, and it may be satisfactory when workpieces are to be machined after heat treating, because natural atmospheres are strongly oxidizing. The heavily oxidized surfaces created by use of this type of atmosphere inhibit cooling rate during quenching.

Products of combustion in direct-fired furnaces automatically provide some atmosphere protection compared with exposure to air. When fuels are mixed with air and burned at the ideal ratio, a condition results wherein minimal reaction with the steel surfaces occurs. This ideal ratio varies for different fuels; for example, in burning natural gas (methane), this ratio is approximately 10 parts air to 1 part gas; whereas for propane, the ideal ratio is 23 parts air to 1 part gas. When an excess of air exists in the mixture, loose scale forms. When the mixture contains an excess of fuel, a tight adherent oxide is formed on the workpiece surfaces. In all instances, a certain amount of water vapor develops in the mixture, which causes decarburization in higher carbon steels. Although the atmosphere protection derived by control of combustion is by no means a "perfect atmosphere," it is inexpensive and is sufficient for many applications.

Exothermic atmospheres are the most widely used prepared atmospheres due to their low cost. There are several modifications of this general type of atmosphere, but rich exothermic gas is produced by combustion of a hydrocarbon fuel such as natural gas or propane with the air/fuel ratio closely controlled. This air/gas mixture is burned in a confined combustion space to maintain a reaction temperature of at least 1800 °F (980 °C) for sufficient time to permit the combustion reaction to reach equilibrium. Heat is obtained directly from combustion, hence the term exothermic. The resultant gas is then cooled to remove part of the water vapor formed by burning and to permit convenient transportation and metering. In this process, the simplified theoretical reaction of methane with air is:

$$CH_4 + 1.25O_2 + 4.75N_2 \rightarrow 0.375CO_2 + 0.625CO + 0.88H_2 + 4.75N_2 + 1.12H_2O + heat$$

where 1 volume of fuel and 6 volumes of air yield 6.63 volumes of product gas mixture, with water vapor removed. In practice, exothermic gas generators are seldom operated with an air/gas ratio lower than about 6.6 to 1 to prevent sooting.

Exothermic atmospheres serve a variety of applications including clean (even bright) annealing and clean hardening of steel, as well as some nonferrous metals. They are not well suited to processing high-carbon steels, unless the workpieces are subjected to stock removal in finishing, because exothermic atmospheres decarburize high-carbon steels.

Endothermic atmospheres are more costly than are exothermic, but they are more flexible than the exothermic types.

Endothermic atmospheres are produced in generators that use air and a hydrocarbon gas as fuel. These two gases are mixed in a controlled ratio, slightly compressed, and then passed into a chamber that is filled with a nickel-bearing catalyst. This chamber has been heated externally to approximately 1900 °F (1040 °C), hence the term endothermic. The gases react in this chamber to form endothermic gas. Figure 15 is a schematic diagram of an endothermic gas generator.

Endothermic atmospheres can be used in virtually all furnace processes that operate above 1400 °F (760 °C). The most common use is as carrier gases in gas carburizing and carbonitriding applications. Because of the wide range of possible carbon equivalencies, however, endothermic atmospheres also are used for bright hardening of steel, for carbon restoration in forgings and bar stock, and for sintering of powder compacts that require reducing atmospheres.

Commercial nitrogen-base atmospheres are used for many heat treating applications (sometimes replacing the endothermic atmospheres).

Commercial nitrogen-base atmosphere systems employed by the

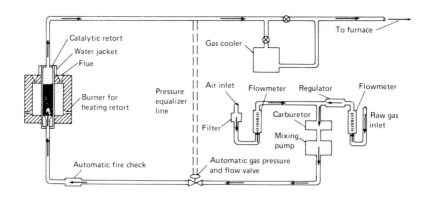

Fig. 15 Schematic flow diagram of an endothermic gas generator. Source: *Metals Handbook*, Vol 4, 9th edition, American Society for Metals, p 397, 1980.

metalworking and heat treating industry use gases and equipment that are common among all applications. In most instances, the major atmosphere component is industrial gas nitrogen, which is supplied to the furnace from a system consisting of a storage tank, vaporizer, and a station controlling pressure and flow rate. The nitrogen serves as a pure, dry, inert gas that provides an efficient purging and blanketing function within the heat treating furnace. The nitrogen stream is often enriched with a reactive component, and the resulting composition and flow rate are determined by the specific furnace design, temperature, and material being heat treated. Although there is similarity in the components of commercial nitrogen-base atmosphere systems, the flexibility of controlling atmosphere composition and flow rate independently over a wide range provides very different end use characteristics. Therefore, the classification of commercial nitrogen-base atmosphere systems is appropriately made according to three major categories of atmosphere function—protection, reactivity, and carbon-control—rather than by gas or equipment components.

The basic components of an industrial gas atmosphere system are illustrated in Fig. 16. Figure 16(a) shows a nitrogen-hydrogen protective atmosphere system used for annealing, brazing, and sintering. Figure 16(b) is a nitrogen-methanol system typical of those used for carburizing or neutral heating. In both instances, there are three basic parts to the commercial-nitrogen system: the storage vessels containing the elemental atmosphere components, the blend panel used to control the flow rate of each constituent gas, and the piping and wiring required for safe operation compatible with the furnace design.

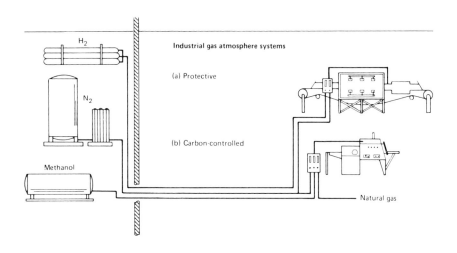

Fig. 16 Schematic presentation of two nitrogen-base atmosphere systems. Source: *Metals Handbook*, Vol 4, 9th edition, American Society for Metals, p 404, 1980.

Nitrogen is the main component of most commercial nitrogen-base systems. Because nitrogen is noncorrosive, special materials of construction are not required, except that they must be suitable for the temperatures of liquid nitrogen. Tanks may be spherical or cylindrical in shape.

Hydrogen is used as a reactive reducing gas for many heat treating atmosphere applications. In commercial nitrogen-base systems, hydrogen is normally blended as a gas with nitrogen to form an atmosphere composition of 90% nitrogen and 10% hydrogen.

Vacuum atmospheres are flexible, in terms of applications, and easily generated. Equipment cost is high, however.

Dissociated ammonia provides a relatiely high-cost prepared furnace atmosphere providing a dry, carbon-free source of reducing gas. Typical composition is 75% hydrogen, 25% nitrogen, less than 300 ppm residual ammonia, and less than -60 °F (-50 °C) dew point.

Principal uses of dissociated-ammonia furnace atmosphere include bright copper and silver brazing; bright heat treating of selected nickel alloys, copper alloys, and carbon steels; bright annealing of electrical components; and as a carrier mixed gas for certain nitriding processes. Dissociated ammonia ($N_2 + 3H_2$) is produced from commercially supplied anhydrous ammonia (NH_3) with an ammonia dissociator. This equipment raises the temperature of ammonia vapor in a catalyst-filled retort to approximately 1650 to 1800 °F (900 to 980 °C). The gas is then cooled for metering and transport as a prepared atmosphere. At these reaction temperatures in the presence of catalyst, ammonia vapor dissociates into separate constituents of hydrogen and nitrogen.

Dry Hydrogen Atmospheres. Commercially available hydrogen is 98 to 99.9% pure. All cylinder hydrogen contains traces of water vapor and oxygen. Methane, nitrogen, carbon monoxide, and carbon dioxide may be present as impurities in very small amounts, depending on the method of manufacture. Hydrogen is produced commercially by a variety of methods, including the electrolysis of water, the catalytic conversion of hydrocarbons, the decomposition of ammonia, and the water gas reaction.

Dry hydrogen is used in the annealing of stainless and low-carbon steels, electrical steels, and several nonferrous metals. It is used also in the sintering of refractory materials such as tungsten carbide and tantalum carbide, in the nickel brazing of stainless steel and heat-resistant alloys, and in copper brazing, direct reduction of metal ores, annealing of metal powders, and the sintering of powder metallurgy compacts.

Argon, as received in tanks (such as used for gas shielded-arc welding), provides an excellent neutral atmosphere. However, due to high cost, its use is confined mainly to heat treating of certain exotic alloys in small furnaces.

Steam Atmospheres. Steam may be used as an atmosphere for scale-free tempering and stress relieving of ferrous metals in the temperature range of

650 to 1200 °F (345 to 650 °C). The steam causes a thin, hard and tenacious blue-black oxide to form on the metal surface. This oxide film, which is about 0.00005 to 0.0003 in. (0.00127 to 0.008 mm) thick, improves certain properties of various metal parts.

Before parts are processed in steam atmospheres, their surfaces must be clean and oxide-free, to permit the formation of a uniform coating. To prevent condensation and rusting, steam should not be admitted until workpiece surfaces are above 212 °F (100 °C). Air must be purged from the furnace before the temperature exceeds 800 °F (425 °C) to prevent the formation of a brown coating instead of the desired blue-black coating.

Temperature Control Systems

Temperature control in heat treating is of paramount importance for maintenance of quality. Process temperature should be controlled to within about ±5 °F (±2.5 °C). Although this close range is sometimes possible, a more practical control range is nearer ±10 °F (±5 °C).

In any temperature control system, three steps must be executed. Before control can be established, the variable must first be "sensed" by some device that responds to changes in the quality or value of the variable. This quantity, or its change, must then be indicated or recorded prior to being controlled. Following the control action, the last step in the sequence is the transmission of the controller output to the "final element," which is a component of the process itself. Final elements relay the output of the controller and cause corrective changes in the process.

Temperature Sensors. As is often the case, one variable is measured then translated, or converted, to another. For example, environmental temperatures actually are measured by expansion or contraction of a column of fluid or of a metal. By means of calibration, these variables are converted to numerical temperature readings.

These simple devices, however, are not suitable for the higher temperatures involved in most heat treating operations. Thermocouples are the most widely used sensors for measuring temperatures of heat treating furnaces. In temperature measurement, however, the same approach is used; that is, one variable is measured then converted to another.

Thermocouples consist of two dissimilar metal wires that are metallurgically homogeneous. They are joined at one end—the measuring or hot junction. The other end, which is connected to the copper wire of the measuring instrument circuitry, is called the reference or cold junction. The electrical signal output in electromotive force (EMF) is proportional to the difference in temperature between the measuring junction (hot) and the reference junction (cold) (Fig. 17a). The signal in millivolts is transmitted to the indicating instrument (Fig. 17b), which translates millivolts to degrees of temperature. The indicating phase of the system must, however, be calibrated to the two dissimilar metals used in the thermocouple.

There are a number of standard metal combinations that may be used for temperature measurement. The most common combinations, along with their letter type designations and applicable temperature ranges, are listed in Table 2. Type **K** is, by far, the most widely used.

To obtain accurate temperature sensing, thermocouples must be placed near the work. In many instances, more than one thermocouple is used within a given furnace.

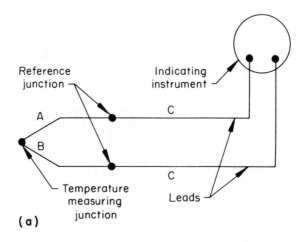

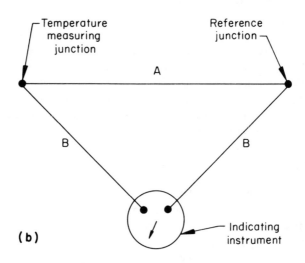

Fig. 17 Principle of the thermocouple. (a) Thermoelectric circuit used for measuring temperature. (b) Simple thermocouple. Source: MEI Course 6, Lesson 12, American Society for Metals, p 3, 1980.

Radiation pyrometers also can be used to sense temperatures in heat treating furnaces. This method uses an extremely accurate "electric eye" to sense temperature changes based on changes in wavelengths. For example, the change of a metal from red hot to white hot as temperature increases corresponds to the change from longer to shorter wavelengths. Wavelength can be sensed without contact with the body whose temperature is being measured.

The amount of radiant energy transmitted from a hot body to a cold body is a function of the difference between the fourth powers of their absolute temperatures. Consequently, sensors that measure the amount of energy (heat) emitted can develop an output signal related to temperature. The sensors provide optical systems that focus the energy received on detectors such as thermocouples, thermopiles, thermistors, or photovoltaic cells.

The hot body may radiate or reflect energy it receives, depending on its characteristics. If it radiates 100% of the energy it receives, it is called a black body and is said to have an emissivity of 1. If it radiates only 50% of the energy, it has an emissivity of 0.5. Consequently, the emissivity of the hot body affects the measurement of its temperature.

Radiation devices are available in various ranges such as 800 to 1800 °F, 1400 to 3000 °F, and 1850 to 4000 °F (425 to 980 °C, 760 to 1650 °C, and 1010 to 2200 °C, respectively). They are also made with varying speeds of response to meet the demands of different applications. For example, detectors having a high speed of response (0.5 s) are useful where the object is moving or where the temperature of the object varies rapidly. Slow response speeds are used for filtering out rapid or erratic changes in temperature such as given by swirling flames. The details of one type of radiation sensor are shown in Fig. 18.

Table 2 Thermocouple types, nominal temperature ranges and material combinations

Source: MEI Course 20, Lesson 8, American Society for Metals, p 5, 1980.

Type	Nominal temperature range °F	°C	Typical thermocouple material (a)
B..................120-3300		50-1815	PLATINUM, 30% RHODIUM-platinum, 6% rhodium
E..................32-1600		0-870	CHROMEL-constantan
J.................. −300-1400		−185-760	IRON-constantan
K..................32-2300		0-1260	NICKEL, CHROMIUM-nickel, aluminum
R..................32-2700		0-1480	PLATINUM, 13% RHODIUM-platinum
S 32-2700		0-1480	PLATINUM, 10% RHODIUM-platinum
T ....−300-700		−185-370	COPPER-constantan
W5 0-4000		−20-2205	TUNGSTEN, 5% RHENIUM-tungsten, 26% rhenium

(a) Upper case letters indicate the positive lead.

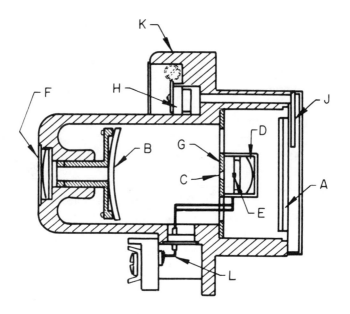

Fig. 18 Total radiation pyrometer. A, window; B, mirror; C, opening to thermopile; D, mirror; E, thermopile; F, lens; G, receiving disk; H, screw and gear adjustment for shutter; J, shutter (adjusted at factory only); K, pyrometer housing; and L, electrical lead wires to temperature indicating instrument. Source: MEI Course 6, Lesson 12, American Society for Metals, p 14, 1980.

Temperature measurement represents the second step in a temperature control system (Ref 10). Measurement instruments measure the output signal of the temperature sensor and convert it to a temperature indication or recording in engineering units. Transmitters are used in some measurement systems to amplify and condition the temperature signal. The accuracy of the measurement depends greatly on the accuracy of the temperature sensor and the connecting lead wire. The accuracy of the measurement instrument is defined in its specifications under referenced conditions for its power supply, ambient conditions (temperature and humidity), electrical noise rejection, and maximum source impedance. The accuracy of the transmitter has similar qualifications.

Measurement instruments are classified by their displays, analog or digital, and whether they are recording or nonrecording types. Analog displays include meters and motor-driven pointers. Analog strip chart or round chart recorders include analog temperature indication. Digital displays are available with or without digital printers. In general, digital equipment is more accurate than analog equipment, but specifications must be checked in each case. The analog versus digital choice in displays and recording usually depends on user preference and specific applications. Typical digital displays are shown in Fig. 19.

Temperature measurement instruments incorporate reference junction compensation in their circuitry for thermocouple measurements and emissivity compensation for noncontact radiation sensors. Reference junction compensation automatically adjusts the measurement depending on the temperature at the junction between the thermocouple wire and the copper wire of the instrument's measuring circuit. Radiation-type temperature measurements require an emissivity compensator. This compensator makes a calibration adjustment by comparing measurement of the same target with an optical pyrometer or calibrated thermocouple. The emissivity compensator is adjusted to make the radiation pyrometer measurement indication agree with the reference calibration instrument.

Temperature control is the third major phase of a temperature control system. A temperature controller must provide sufficient energy to satisfy process requirements, even though operating conditions vary. Variations include changes in process load, fuel characteristics, and ambient temperature. Thus, controller requirements are more stringent when process requirements are demanding and especially when operating conditions vary significantly.

In operation, the controller set point that represents the desired temperature is compared with the process or actual temperature. The stability of the controller and its sensitivity to the difference between desired and

Fig. 19 Digital programmer for set point and logic control. Source: MEI Course 20, Lesson 8, American Society for Metals, p 15, 1980.

actual temperatures are critical. Based on this comparison, the controller regulates the energy flow to the process.

The two basic types of control are the two-position, or on-off type, and the proportioning, or modulating type. These two basic types exist in many variations.

Generally, proportioning, or modulating, types are preferred as opposed to the on-off types, because the proportioning types (regardless of the source of energy) permit closer temperature control.

Chapter 5

Heat Treating
of Carbon Steels

As discussed in Chapter 2, carbon steels are not strictly alloys of iron and carbon; manganese, silicon, phosphorus, and sulfur usually are present in small amounts. This condition has always led to some confusion regarding what carbon steels are, as well as what they are not.

Much study and effort has gone into developing a definition of carbon steels, as well as a list of compositions that are universally accepted by steel producers and fabricators. These efforts are summarized in Tables 1 to 6. The development and standardization of carbon steel compositions has been established by the American Iron and Steel Institute (AISI) and the Society of Automotive Engineers (SAE). These standards for carbon steels cover composition only. Hundreds of standards or codes that are established by other societies or agencies exist which cover steels or other alloys. Frequently, these standards place primary emphasis on mechanical properties rather than on composition. For this reason, it becomes difficult and sometimes dangerous to translate one specification to another. No attempt is made in this book to deal with compositions other than those established by AISI-SAE, except for the newer UNS system described below.

Unified Numbering System

The standard carbon and alloy grades established by AISI or SAE have now been assigned designations in the Unified Numbering System (UNS) by the American Society for Testing and Materials (ASTM E527) and the Society of Automotive Engineers (SAE J1086). In the composition tables in this chapter, UNS numbers are listed along with their corresponding AISI-SAE numbers where available.

The UNS number consists of a single letter prefix followed by five numerals. The prefix letter G indicates standard grades of carbon or alloy steels, while the prefix letter H indicates standard grades which meet certain hardenability limits. The first four digits of the UNS designations usually correspond to the standard AISI-SAE designations, while the last digit (other than zero) denotes some additional composition requirement such as lead or boron. The digit is sometimes a 6, which is used to designate steels which are made by the basic electric furnace with special practices. The AISI alloys steels in Chapter 6 are also covered by the UNS system.

What Are Carbon Steels?

By the AISI classification, a steel is considered to be a carbon steel when:

- No minimum content is specified or required for aluminum, boron, chromium, cobalt, columbium, molybdenum, nickel, titanium, tungsten, vanadium, zirconium, or any other element added to obtain a desired alloying effect.
- The specified minimum for copper does not exceed 0.40%.
- The maximum specified content does not exceed the following limits: manganese, 1.65%; silicon, 0.60%; and copper, 0.60%.

Steels are sold to definite chemical limits, and each grade of plain carbon steel is assigned a code number that specifies its chemical composition. There are several coding systems, but the AISI system is the most comprehensive and widely used. It uses a code of four digits for all carbon steels. The first two always indicate the general type of steel. For instance, consider a steel identified as AISI 1040 grade. The digits 10 indicate it to be a plain carbon steel with a manganese content of no greater than 1.0%. The next two digits, 40, indicate the mean carbon content, in this case 0.40%. Thus, when the code 1040 is seen, the reader knows the steel is a plain carbon grade containing 0.40% C plus or minus a few points. The steel described above is the first entry in Table 1. The 10XX means that it is a plain carbon steel—the XX indicates that the carbon content may vary.

Table 1 Types and approximate percentages of identifying elements in standard carbon and alloy steels

Source: Heat Treater's Guide: Standard Practices and Procedures for Steel, American Society for Metals, p 15, 1982.

Series designation	Description
Carbon steels	
10XX	Nonresulfurized, 1.00 manganese maximum
11XX	Resulfurized
12XX	Rephosphorized and resulfurized
15XX	Nonresulfurized, over 1.00 manganese maximum

Approximately 20 different compositions of the 10XX group are listed in Table 2. The complete list contains over 40 compositions, but the 20 that are omitted are interspersed within the group shown and vary only slightly in carbon and/or manganese content. The large number of compositions that vary minimally are designated as such for the convenience of the steel producer and are of little interest to the heat treater. Steels designated as 1012, 1017, 1037, 1044, 1059, and 1086 are examples of compositions purposely omitted from Table 2. Some of these grades are produced only for wire rods and wire. From Table 2, it is evident that the entire carbon range of 0.08 to 0.95% is covered and in relatively small increments of carbon.

All of the grades listed in Table 2 are considered as very low hardenability steels; none carry the "H" suffix, which is used to denote a specific hardenability that is guaranteed by the steel producer. Steel 1008 is incapable of developing much hardness even when heated above its transformation temperature and drastically quenched, because there is not enough carbon present. On the other hand, 1095 steel can develop a hardness of 66 HRC when very thin sections are cooled quickly, or high hardnesses can be

Table 2 Compositions of standard nonresulfurized carbon steels (1.0% Mn)
Source: Heat Treater's Guide: Standard Practices and Procedures for Steel, American Society for Metals, p 16, 1982.

Steel designation AISI or SAE	UNS No.	C	Mn	P max	S max
1008	G10080	0.10 max	0.30-0.50	0.040	0.050
1012	G10120	0.10-0.15	0.30-0.60	0.040	0.050
1015	G10150	0.13-0.18	0.30-0.60	0.040	0.050
1018	G10180	0.15-0.20	0.60-0.90	0.040	0.050
1020	G10200	0.18-0.23	0.30-0.60	0.040	0.050
1022	G10220	0.18-0.23	0.70-1.00	0.040	0.050
1025	G10250	0.22-0.28	0.30-0.60	0.040	0.050
1029	G10290	0.25-0.31	0.60-0.90	0.040	0.050
1030	G10300	0.28-0.34	0.60-0.90	0.040	0.050
1035	G10350	0.32-0.38	0.60-0.90	0.040	0.050
1038	G10380	0.35-0.42	0.60-0.90	0.040	0.050
1040	G10400	0.37-0.44	0.60-0.90	0.040	0.050
1045	G10450	0.43-0.50	0.60-0.90	0.040	0.050
1050	G10500	0.48-0.55	0.60-0.90	0.040	0.050
1055	G10550	0.50-0.60	0.60-0.90	0.040	0.050
1060	G10600	0.55-0.65	0.60-0.90	0.040	0.050
1070	G10700	0.65-0.75	0.60-0.90	0.040	0.050
1080	G10800	0.75-0.88	0.60-0.90	0.040	0.050
1084	G10840	0.80-0.93	0.60-0.90	0.040	0.050
1090	G10900	0.85-0.98	0.60-0.90	0.040	0.050
1095	G10950	0.90-1.03	0.30-0.50	0.040	0.050

produced in the surface layers by drastic quenching. This practice is sometimes termed "shell hardening" because a very hard shell is produced on the outside of the workpiece, while the subsurface layers are relatively soft because they were not cooled fast enough to exceed the critical cooling rate. This condition is often useful in practical applications.

To provide higher hardenability, additional groups of carbon steels have been developed—higher manganese grades and/or boron treated grades.

Compositions of these special grades, which are considered carbon steels, are listed in Tables 3 and 4.

Higher Manganese Carbon Steels

Because of differences in properties, and in particular their response to heat treatment, the higher manganese carbon steels have been removed from the 10XX series and are assigned to the 15XX series. These steels are marketed as carbon steels, but because of their higher manganese content, they respond to heat treatment like some alloy steels. The same method of designation prevails for the 15XX series; that is, the digits 15 denote higher manganese (up to 1.65%), and the last two digits designate the mean carbon content. For example, steel 1551 has a carbon range of 0.45 to 0.56%.

In carbon steels, for each element specified, there is a permissible range of composition, or a maximum limit, rather than a single specified value. This is because it is difficult to melt to the exact desired chemical percentage of each element; some leeway must be allowed for practical purposes. Manganese, phosphorus, and sulfur appear in the composition of every commercial steel either as residual elements or alloys. In the case of manganese, the amounts present usually require that this element be purposely added. Phosphorus, sulfur, and silicon are useful in their own right in special applications.

Table 3 Compositions of standard nonresulfurized carbon steels (over 1.0% Mn)

Source: Heat Treater's Guide: Standard Practices and Procedures for Steel, American Society for Metals, p 16, 1982.

Steel designation AISI or SAE	UNS No.	Chemical composition, %			
		C	Mn	P max	S max
1513 .... G15130		0.10-0.16	1.10-1.40	0.040	0.050
1522 .... G15220		0.18-0.24	1.10-1.40	0.040	0.050
1524 .... G15240		0.19-0.25	1.35-1.65	0.040	0.050
1526 .... G15260		0.22-0.29	1.10-1.40	0.040	0.050
1527 .... G15270		0.22-0.29	1.20-1.50	0.040	0.050
1541 .... G15410		0.36-0.44	1.35-1.65	0.040	0.050
1548 .... G15480		0.44-0.52	1.10-1.40	0.040	0.050
1551 .... G15510		0.45-0.56	0.85-1.15	0.040	0.050
1552 .... G15520		0.47-0.55	1.20-1.50	0.040	0.050
1561 .... G15610		0.55-0.65	0.75-1.05	0.040	0.050
1566 .... G15660		0.60-0.71	0.85-1.15	0.040	0.050

Generally, however, they are undesirable and are considered impurities that have been carried along in the materials used to make up the steel. Because their complete removal is quite expensive, their presence is reduced to some nominal amount (0.04% maximum or less for phosphorus and sulfur), which renders them practically ineffective as far as the properties of the steel are concerned. They are then carried along in this reduced amount rather than eliminated entirely.

In studying Table 3, note that none of the eleven 15XX steels have the "H" suffix; therefore, no specific hardenability is guaranteed. Although the manganese ranges are generally higher comnpared with the 10XX steels (Table 2), there is a substantial variation in the manganese ranges for the steels shown in Table 3. In two instances (1524 and 1541), the upper boundaries of the manganese ranges reach the maximum limits that allow them to be considered as carbon steels. This variation in manganese content causes a substantial difference in hardenability and thus in response to heat treatment.

Carbon H-Steels

A few carbon steels are available as H-grades. Compositions for six of these steels are listed in Table 4—two 10XX and four 15XX grades. In Table 4, note that 1524H and 1541H permit manganese contents up to 1.75%. These represent exceptions to the general description of carbon steels given above. As a rule, when the manganese content exceeds 1.65%, a steel is considered an alloy steel.

For hardenability bands of all H-steels (see Ref 7), the data presented in Fig. 1 illustrate the positive effect of manganese on hardenability.

The bottom graph and table in Fig. 1 show the hardenability for 1038H steel with a manganese range of 0.60 to 0.90%. Maximum hardness at the quenched end of the end-quench bar (see Chapter 3) approaches 60 HRC, but drops drastically away from the end (commonly described as a steep hardenability curve). Therefore, this steel is considered to be very low in hardenability.

Referring to the middle graph and table of Fig. 1, the effect of manganese on hardenability is readily evident for 1541H. Maximum hardness shown for both boundaries of the band shows a slightly higher initial hardness for 1541H compared with 1038H, but this is due to the slightly higher carbon range for 1541H. The less steep curves that characterize the 1541H grade are the result of higher manganese content.

Boron-Treated Carbon Steels

Another approach to increasing the hardenability of carbon steels is by treatment with boron. This approach has become quite popular with fabricators as a means of obtaining greater hardenability without using more expensive alloy steels.

Because of the very small amount of boron used (0.0005 to 0.003%), steels containing boron to increase their hardenability are usually termed as boron treated, rather than regarding boron as an alloying element. Sometimes, these alloys are called "needled steels." Carbon steels that contain boron are identified by inserting the letter "B" between the second and third digits of the AISI number. For instance, a boron treated 1541 steel would be written as 15B41.

Table 4 Compositions of standard carbon H-steels and standard carbon boron H-steels

Source: Heat Treater's Guide: Standard Practices and Procedures for Steel, American Society for Metals, p 16, 1982.

Steel designation AISI or SAE	UNS No.	C	Mn	P max	S max	Si
Standard carbon H-steels						
1038H	H10380	0.34-0.43	0.50-1.00	0.040	0.050	0.15-0.30
1045H	H10450	0.42-0.51	0.50-1.00	0.040	0.050	0.15-0.30
1522H	H15220	0.17-0.25	1.00-1.50	0.040	0.050	0.15-0.30
1524H	H15240	0.18-0.26	1.25-1.75(a)	0.040	0.050	0.15-0.30
1526H	H15260	0.21-0.30	1.00-1.50	0.040	0.050	0.15-0.30
1541H	H15410	0.35-0.45	1.25-1.75(a)	0.040	0.050	0.15-0.30
Standard carbon boron H-steels						
15B21H	H15211	0.17-0.24	0.70-1.20	0.040	0.050	0.15-0.30
15B35H	H15351	0.31-0.39	0.70-1.20	0.040	0.050	0.15-0.30
15B37H	H15371	0.30-0.39	1.00-1.50	0.040	0.050	0.15-0.30
15B41H	H15411	0.35-0.45	1.25-1.75(a)	0.040	0.050	0.15-0.30
15B48H	H15481	0.43-0.53	1.00-1.50	0.040	0.050	0.15-0.30
15B62H	H15621	0.54-0.67	1.00-1.50	0.040	0.050	0.40-0.60

(a) Standard AISI-SAE H-steels with 1.75 manganese maximum are classified as carbon steels

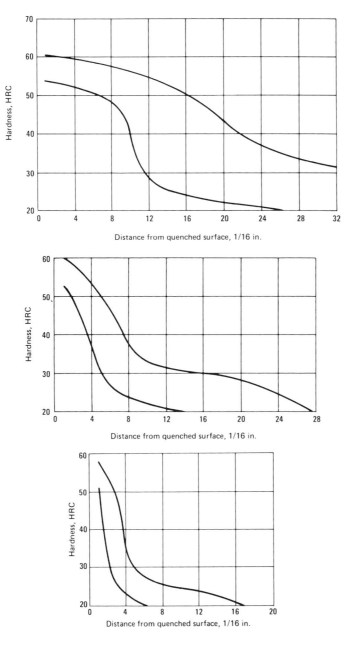

Fig. 1 Lower graph features the low hardenability of a conventional carbon steel; center graph shows effect of manganese on hardenability; upper graph shows effect of manganese and boron on hardenability (see text). Source: *Heat Treater's Guide: Standard Practices and Procedures for Steel,* American Society for Metals, p 51, 104, 107, 1982.

Compositions for six standard grades of boron-treated carbon steels are listed in Table 4. However, other grades of carbon steels are available as boron-treated grades, usually by special order.

The profound effect of boron on the hardenability is demonstrated in Fig. 1. These two steels shown are essentially the same in composition except for the boron treatment. The effect of boron is quite evident on 15B41H. The precise mechanism of how boron (especially in such small amounts) accomplishes what it does is not entirely clear. However, it is well known that it slows the critical cooling rate; thus, it follows that hardenability increases.

Free-Machining Carbon Steels

For the carbon steels discussed above, sulfur and phosphorus were generally regarded as impurities; thus, it was desirable to keep their contents as low as possible. Free-machining carbon steels contain controlled amounts of these elements to improve machinability.

There are presently two different series of free-machining carbon steels. First is the series assigned AISI 11XX, referred to as resulfurized carbon steels. The carbon content ranges vary from 0.08 to 0.13% C for AISI 1110 to 0.48 to 0.55% C for AISI 1151. Several members of this series also contain a relatively high manganese content (as high as 1.65% for 1144), and sulfur contents as high as 0.33%. Phosphorus content for this series is, however, restricted to a maximum of 0.040%. Compositions for the standard resulfurized grades are given in Table 5.

A second series of free-machining carbon steels is identified as rephosphorized and resulfurized carbon steels and is coded as AISI 12XX.

All of the 12XX grades are low in carbon content (0.15% maximum). Also, the maximum manganese content is generally lower compared with the 11XX series. Sulfur content is also high in the 12XX steels (0.35% maximum for two grades), but the outstanding difference between the 11XX and 12XX series is that the steels of the latter series have been rephosphorized as well as resulfurized, for better machinability, than can be provided by resulfurization alone. Compositions of the 12XX series are presented in Table 6.

Leaded Steels. The last composition shown in Table 6 also contains lead as indicated by the letter "L" between the second and third digits of the designation (12L14). While this is the only standard grade that contains lead and controlled amounts of sulfur and/or phosphorus, almost any of the free-machining grades, as well as many of the nonfree-machining grades (10XX and 15XX series), are available by special order with lead additions. Although lead additions do not provide the degree of improved machinability that can be provided by resulfurization and/or rephosphorization, neither is there as great a sacrifice of mechanical properties for lead additions compared with sulfur and/or phosphorus additions.

Table 5 Compositions of standard resulfurized carbon steels

Source: Heat Treater's Guide: Standard Practices and Procedures for Steel, American Society for Metals, p 15, 1982.

Steel designation AISI or SAE	UNS No.	Chemical composition, % C	Mn	P max	S
1110 .... G11100		0.08-0.13	0.30-0.60	0.040	0.08-0.13
1117 .... G11170		0.14-0.20	1.00-1.30	0.040	0.08-0.13
1118 .... G11180		0.14-0.20	1.30-1.60	0.040	0.08-0.13
1137 .... G11370		0.32-0.39	1.35-1.65	0.040	0.08-0.13
1139 .... G11390		0.35-0.43	1.35-1.65	0.040	0.13-0.20
1140 .... G11400		0.37-0.44	0.70-1.00	0.040	0.08-0.13
1141 .... G11410		0.37-0.45	1.35-1.65	0.040	0.08-0.13
1144 .... G11440		0.40-0.48	1.35-1.65	0.040	0.24-0.33
1146 .... G11460		0.42-0.49	0.70-1.00	0.040	0.08-0.13
1151 .... G11510		0.48-0.55	0.70-1.00	0.040	0.08-0.13

Effects of Free-Machining Additives on Properties and Heat Treating Procedures

To say the least, the free-machining additives are not beneficial to the mechanical properties of the workpieces into which these steels are made or to heat treating procedures. Free-cutting additives create a dispersion of tiny voids in the steel, which is undesirable. Therefore, in using any free-machining steel, the decision must be made as to whether the impairment of mechanical properties can be tolerated to gain better machinability, which is the sole benefit.

Effects of Sulfur. Additions of sulfur probably disturb the mechanical properties more than either of the other two additives. Figure 2 shows the polished and unetched surface of a specimen of 1110 steel (see Table 5). Sulfur combines with iron and/or manganese to form the prominent inclusions. Because inclusions have no strength, each one creates a tiny void. They serve as chip breakers in machining applications and prevent buildup on edges of cutting tools, sometimes eliminating the need for cutting fluids. Inclusions are "built in" and are thus more effective in this area. Resulfurization has various effects on properties of finished workpieces—notably on impact strength and transverse tensile strength (opposite to rolling direction).

As a rule, the effects of sulfur on heat treating procedures is minimal; that is, the procedures for heat treating a resulfurized steel versus its nonresulfurized counterpart (same carbon content) are generally the same. There is one factor, however, that can be misleading to the heat treater if not clearly understood. In the higher sulfur grades such as 1144, sulfur combines with manganese, forming manganese sulfide, which impoverishes the matrix of its manganese and lowers hardenability. For example, 1144 steel may contain up to 1.65% Mn, which suggests that its hardenability may be about

Table 6 Compositions of standard rephosphorized and resulfurized carbon steels
Source: Heat Treater's Guide: Standard Practices and Procedures for Steel, p 18, 1982.

Steel designation AISI or SAE	UNS No.	Chemical composition, %				
		C	Mn	P	S	Pb
1211	G12110	0.13 max	0.60-0.90	0.07-0.12	0.10-0.15	...
1212	G12120	0.13 max	0.70-1.00	0.07-0.12	0.16-0.23	...
1213	G12130	0.13 max	0.70-1.00	0.07-0.12	0.24-0.33	...
1215	G12150	0.09 max	0.75-1.05	0.04-0.09	0.26-0.35	...
12L14	G12144	0.15 max	0.85-1.15	0.04-0.09	0.26-0.35	0.15-0.35

equal to that of 1541. However, this is not necessarily true; 1144 has a very high sulfur range (0.24 to 0.33%) so that some of the manganese may be tied up by the sulfur.

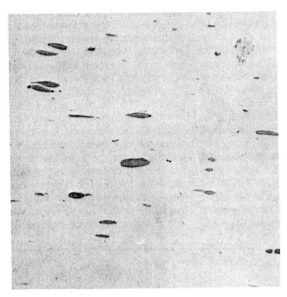

Fig. 2 Sulfide inclusions (gray areas) in grade 1110 free-machining steel. Source: *Metals Handbook*, Vol 1, 9th edition, American Society for Metals, p 577, 1977.

Effects of phosphorus are similar to those described above for sulfur. Because phosphorus is not added in great amounts, its effect on mechanical properties is minimal and its effect on heat treating procedures is essentially nil.

Effects of Lead. Lead may be added to any of the carbon steels, as well as to alloy steels (see Chapter 6), to provide improved machinability. However, the amount of improvement provided by the addition of 0.15 to 0.35% Pb is not nearly as great compared with sulfur and/ or phosphorus additions. Likewise, lead additions are far less likely to impair mechanical properties compared with other additives, notably sulfur. Leaded steels are produced by adding fine lead shot to a stream of molten steel as it enters the ingot mold. Because lead is totally insoluble in steel, it supposedly remains as a very fine dispersion of almost submicroscopic particles in the solidified steel, if the leading procedure is properly performed. Lead segregations can and sometimes do occur, which can cause serious difficulty in heat treatment. Almost every heat treater who has treated quantities of leaded steel parts has had the disheartening experience of finding holes or porosity in the heat treated workpieces which were caused by segregations of lead that melted out during heat treatment.

General Precautions. None of the carbon steels that contain free-machining additives, and especially the resulfurized grades, should be considered for applications that require any appreciable amount of cold forming. Likewise, while free-machining steels are not totally unforgeable, they are prone to split in forging. Consequently, they should be considered only for forgings of very mild severity.

Classification for Heat Treatment

To simplify consideration of the treatments for various applications, the steels in this discussion are classified as follows: Group I, 0.08 to 0.25% C; Group II, 0.30 to 0.50% C; and Group III, 0.55 to 0.95% C. A relatively few steels, such as 1025 and 1029, can be assigned to more than one group, depending on their precise carbon content.

Group I (0.08 to 0.25% C). The three principal types of heat treatment used on these low-carbon steels are (1) process treating of material to prepare it for subsequent operations; (2) treating of finished parts to improve mechanical properties; and (3) case hardening, notably by carburizing or carbonitriding to develop a hard, wear-resistant surface. It is often necessary to process anneal drawn products between operations, thus relieving work strains in order to permit further working (see Fig. 3). This operation normally is carried out at temperatures between the recrystallization temperature and the lower transformation temperatures. The effect is to soften by recrystallization of ferrite. It is desirable to keep the recrystallized grain size relatively fine. This is promoted by rapid heating and short holding time at temperature. A similar practice may be used in the treatment of low-carbon, cold headed bolts made from cold drawn wire. Sometimes, the strains introduced by cold working so weaken the heads that they break through the most severely worked portion under slight additional strain. Process annealing is used to overcome this condition. Stress relieving at about 1000 °F (540 °C) is more effective than annealing in retaining the normal mechanical properties of the shank of the cold headed bolt.

Heat treating frequently is used to improve machinability. The generally poor machinability of the low-carbon steels, except those containing sulfur or other additive elements, results principally from the fact that the proportion of free ferrite to carbide is high. This situation can be modified by putting the carbide into its most voluminous form—pearlite— and dispersing fine particles of this pearlite evenly throughout the ferrite mass. Normalizing commonly is used with success, but best results are obtained by quenching the steel in oil from 1500 to 1600 °F (815 to 870 °C). With the exception of steels containing a carbon content approaching 0.25%, little or no martensite is formed, and the parts do not require tempering.

Group II (0.30 to 0.50% C). Because of the higher carbon content, quenching and tempering become increasingly important when steels of this group are considered. They are the most versatile of the carbon steels, because

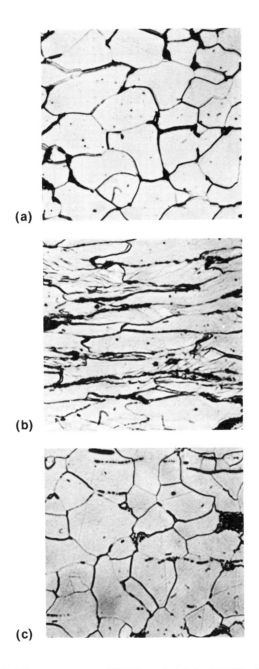

Fig. 3 Microstructure of 1008 steel. (a) At 1000X after slight cold reduction (b) Same steel but after a 60% cold reduction (c) Same steel but after process annealing at 1100 °F (595 °C). Source: *Metals Handbook*, Vol 7, 8th edition, American Society for Metals, p 9, 11, 1972.

their hardenability can be varied over a wide range by suitable controls. In this group of steels, there is a continuous change from water-hardening to oil-hardening types. Hardenability is very sensitive to changes in chemical composition, particularly to the content of manganese, silicon, and residual elements, as well as grain size. These steels are also very sensitive to changes in section size.

The medium-carbon steels should be either normalized or annealed before hardening in order to obtain the best mechanical properties after hardening and tempering. Parts made from bar stock are frequently given no treatment prior to hardening (the prior treatment having been performed at the steel mill), but it is common practice to normalize or anneal forgings.

These steels, whether hot finished or cold finished, machine reasonably well in bar stock form and are machined as received, except in the higher carbon grades and small sizes that require annealing to reduce the as-received hardness. Forgings usually are normalized to improve machinability over that encountered with the fully annealed structure.

In hardening, the selection of quenching medium varies with steel composition, design of the part, hardenability of the steel, and the hardness desired in the finished part. Water is the quenching medium most commonly used, because it is best known and is usually least expensive and easiest to install. Caustic soda solution (5 to 10% NaOH) is used in some instances with improved results. It is faster than water and may produce better mechanical properties in all but light sections. It is hazardous, however, and operators must be protected against contact with it. Salt solutions (brine) are often successfully used. They are not dangerous to operators, but their corrosive action on iron or steel parts or equipment is potentially serious. When the section is light or the properties required after heat treatment are not very high, oil quenching often is used. Finally, medium-carbon steels are readily case hardened by flame or induction hardening (more detailed quenching data are given in Chapter 4).

Group III (0.55 to 0.95% C). Forged parts made of these steels should be annealed for several reasons. Refinement of the forged structure is important in producing a high-quality, hardened product. The parts come from the forging operation too hard for cold trimming of the flash or for any machining operations. Ordinary annealing practice, followed by furnace cooling to about 1100 °F (595 °C) is satisfactory for most parts.

Hardening by conventional quenching is used on most parts made from steels in this group. However, special techniques are required at times. Both oil and water quenching are used; water is used for heavy sections.

Austempering commonly is used for processing these high-carbon grades; see the description of austempering at the end of this chapter.

Tools with cutting edges are sometimes heated in liquid baths to the lowest temperatures at which the part can be hardened and are then quenched in brine. The fast heating of the liquid bath plus the low temperature fail to put all of the available carbon into solution. As a result, the cutting edge consists of martensite containing less carbon than indicated by the chemical composition of the steel and containing many embedded particles of cementite. In this condition, the tool is at its maximum toughness relative to its hardness, and the embedded carbides promote long life of the cutting edge. Final hardness is 55 to 60 HRC. Steels in this group are also commonly hardened by flame or induction methods (see Chapter 11).

Heat Treating Procedures for Specific Grades of Carbon Steels

Detailed heat treating procedures are presented below for a representative group of carbon steels beginning with the very low carbon deep-drawing grade 1008. For greater details or precise procedures on specific steels that are not discussed in this section, see Ref 7.

1008 — Recommended Heat Treating Practice

The most common heat treatment applied to this steel is process annealing, which consists of heating to approximately 1100 °F (600 °C), which recrystallizes cold worked structures. This process frequently is used as an intermediate anneal prior to further cold working (see Fig. 3). Note in Fig. 3(a) that the grains are generally symmetrical because the steel has been subjected to only a slight amount of cold reduction. The low carbon content of 1008 is indicated by the small amount of pearlite (black areas). Figure 3(b) shows the same steel after severe cold reduction; note how the grains have become flattened and elongated. This is accompanied by a marked increase in hardness caused by cold working. Restoration of the low hardness and initial grain structure is accomplished by process annealing at 1100 °F (595 °C), as shown in Fig. 3(c). This treatment, and the results, are representative of the low-carbon grades that are used for cold forming applications.

Case Hardening. Hard, wear-resistant surfaces can be obtained on parts by carbonitriding (see Chapter 10). Depth of case developed depends on time and temperature. Carbonitride at 1400 to 1600 °F (760 to 870 °C). Atmosphere usually is comprised of an enriched carrier gas from an endothermic generator, plus about 10% anhydrous ammonia. Case depths usually range from about 0.003 to 0.010 in. (0.08 to 0.25 mm). Maximum surface hardness usually is obtained by oil quenching directly from the carbonitriding temperature. Cyanide or cyanide-free liquid salt baths may also be used to develop cases essentially the same as those obtained in the gas process. Temperatures, time at temperature, and quenching practice are also approximately the same.

Although the vast majority of parts case hardened by carbonitriding are not tempered, they can be rendered less brittle by tempering at 300 to 400 °F (150 to 205 °C) without appreciable loss of surface hardness.

1020 — Recommended Heat Treating Practice

Normalizing. Heat to 1700 °F (925 °C). Air cool.

Annealing. Heat to 1600 °F (870 °C). Cool slowly, preferably in furnace.

Hardening. Can be case hardened by any one of several processes, which range from light case hardening, such as carbonitriding and the others described for grade 1008, to deeper case carburizing in gas, solid, or liquid mediums. Most carburizing is done in a gaseous mixture of methane combined with one of several carrier gases, using the temperature range of 1600 to 1750 °F (870 to 955 °C). Carburize for desired case depth with a 0.90 carbon potential. Case depth achieved is always a function of time and temperature. For most furnaces, a temperature of 1750 °F (955 °C) approaches the practical maximum without causing excessive deterioration in the furnace. With the advent of vacuum carburizing, temperatures up to 2000 °F (1095 °C) can be used to develop a given case depth in about one half the time required at the more conventional temperature of 1700 °F (925 °C) (see Chapter 10).

Hardening after carburizing is usually achieved by quenching directly into water or brine from the carburizing temperature. After the desired carburizing cycle has been completed, the furnace temperature can be decreased or a lower temperature zone can be used for a continuous furnace to 1550 °F (845 °C) for a diffusion cycle. Quenching into water or brine then tempering at 300 °F (150 °C) follows.

1035 — Recommended Heat Treating Practice

Normalizing, if required, is accomplished by heating to 1675 °F (915 °C) and cooling in still air.

Annealing. Heat to 1600 °F (870 °C). Furnace cool at a rate not exceeding 50 °F (28 °C) per hour to 1200 °F (650 °C).

Hardening. Austenitizing at 1575 °F (855 °C). Quench in water or brine, except for sections under ¼ in. (6.35 mm), which may be oil quenched.

Tempering. As-quenched hardness should be approximately 45 HRC. Hardness can be adjusted downward by tempering (see curve in Fig. 4).

1045, 1045H — Recommended Heat Treating Practice

Normalizing. Heat to 1650 °F (900 °C). Cool in air.

Annealing. Heat to 1550 °F (845 °C). Cool in furnace at a rate not exceeding 50 °F (28 °C) per hour to 1200 °F (650 °C).

Hardening. Austenitize at 1550 °F (845 °C). Quench in water or brine. Oil quench sections under ¼ in. (6.35 mm) thick.

Tempering After Hardening. Hardness of at least 55 HRC, if properly austenitized and quenched. Hardness can be adjusted by tempering (see curve in Fig. 4).

This steel and others having a carbon content within the range of approximately 0.35 to 0.55% usually are normalized followed by tempering.

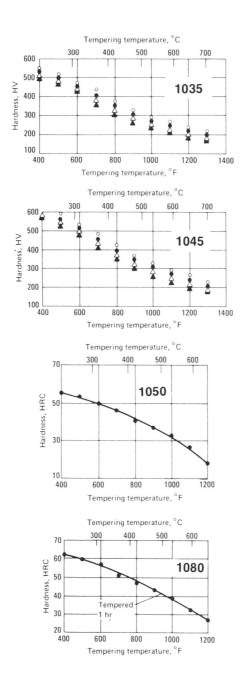

Fig. 4 Tempering curves for several of the higher carbon 10XX series carbon steels. Specimens, 1/8 to 1/4 in. (3.18 to 6.35 mm) thick. O : 10 min. ● : 1 hr. △ : 4 hr. ▲ :24 hr. **1035**: Tempered. **1045**: Quenched. **1050** and **1080**: Quenched. Tempered 1 hr.

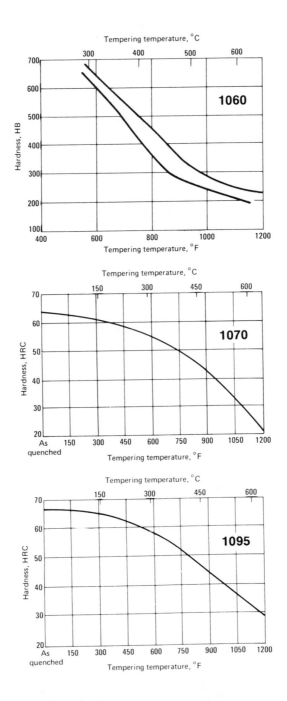

Fig. 4 (continued) **1060**: Specimen, 3/16 to 4-1/8 in. (4.673 to 104.775 mm) thick. Quenched. **1070** and **1095**: Represent an average based on a fully quenched surface.

For thick sections, it is obviously impossible to exceed the critical cooling rate to form martensite. Thus, a compromise is to transform the microstructure by normalizing to fine pearlite (see TTT curve in Chapter 2, also Fig. 5 in this chapter). The normalizing operation is then followed by tempering to relieve stresses and develop the required hardness and/or strength.

1050 — Recommended Heat Treating Practice

Normalizing. Heat to 1650 °F (900 °C). Cool in air.

Annealing. Heat to 1525 °F (830 °C). Cool in furnace at a rate not exceeding 50 °F (28 °C) per hour to 1200 °F (650 °C).

Hardening. Heat to 1525 °F (830 °C). Quench in water or brine. Sections less than ¼ in. (6.35 mm) thick may be fully hardened by oil quenching. Normalize and temper as described for 1045.

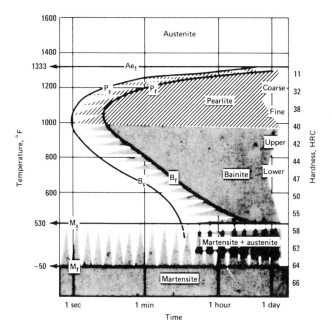

Fig. 5 Complete isothermal transformation diagram for 0.80% carbon steel. All of the transformation products are named. Bainite transformation takes place isothermally between 530 and about 975 °F (275 and about 525 °C). If austenite is rapidly cooled from above A₁ past the nose of the curve and to temperatures below 530 °F (275 °C), M_{51} martensite starts to form. As long as cooling continues, more martensite forms. Transformation of austenite to martensite is not complete until M_1 is reached. Source: *Heat Treater's Guide: Standard Practices and Procedures for Steel*, American Society for Metals, p 11, 1982.

Tempering. As-quenched hardness of 58 to 60 HRC can be adjusted downward by the proper tempering temperature as shown by the curve presented in Fig. 4.

1060 — Recommended Heat Treating Practice

Normalizing. Heat to 1625 °F (885 °C). Cool in air.

Annealing. Heat to 1525 °F (830 °C). Furnace cool to 1200 °F (650 °C) at a rate not exceeding 50 °F (28 °C) per hour.

Hardening. Heat to 1500 °F (815 °C). Quench in water or brine. Oil quench sections under ¼ in. (6.35 mm) thick.

Tempering. As-quenched hardness from 62 to 65 HRC. This maximum hardness can be adjusted downward by proper tempering temperature (see curve in Fig. 4).

Austempering. Thin sections (typically springs) are austempered. Results in a bainitic structure and hardness of approximately 46 to 52 HRC. Austenitize at 1500 °F (815 °C). Quench in molten salt bath at 600 °F (315 °C). Hold at temperature for at least 1 h. Air cool. No tempering required. A more detailed discussion of the austempering practice is given at the end of this chapter.

1070 — Recommended Heat Treatment

Normalizing. Heat to 1625 °F (885 °C). Cool in air.

Annealing. Heat to 1525 °F (830 °C). Furnace cool to 1200 °F (650 °C) at a rate not exceeding 50 °F (28 °C) per hour.

Hardening. Heat to 1500 °F (815 °C). Quench in water or brine. Oil quench sections under ¼ in. (6.35 mm) thick.

Tempering. As-quenched hardness of approximately 65 HRC. Hardness can be adjusted downward by proper tempering as demonstrated by the tempering curve in Fig. 4.

Austempering is also a practical treatment for 1070, using the same practice shown for 1060.

1080 — Recommended Heat Treating Practice

Normalizing. Heat to 1600 °F (870 °C). Cool in air.

Annealing. Heat to 1500 °F (815 °C). Furnace cool to 1200 °F (650 °C) at a rate not exceeding 50 °F (28 °C) per hour.

Hardening. Heat to 1500 °F (815 °C). Quench in water or brine. Oil quench sections under ¼ in. (6.35 mm) thick.

Tempering. As-quenched hardness of approximately 65 HRC. Hardness can be adjusted downward by proper tempering (see curve in Fig. 4).

Austempering. Grade 1080 is frequently subjected to austempering because of its wide usage, namely for small, thin flat springs. The same practice as outlined for 1060 is applicable for 1080.

1095 — Recommended Heat Treating Practice

Normalizing. Heat to 1575 °F (855 °C). Cool in air.

Annealing. As is generally true for all high carbon steels, bar stock supplied by mills in spheroidized condition. Annealed with structure of fine spheroidal carbides in ferrite matrix. When parts are machined from bars in this condition, no normalizing or annealing required. Forgings should always be normalized. Anneal by heating at 1475 °F (800 °C). Soak thoroughly. Furnace cool to 1200 °F (650 °C) at a rate not exceeding 50 °F (28 °C) per hour. From 1200 °F (650 °C) to ambient temperature, cooling rate is not critical. This relatively simple annealing process will provide predominately spheroidized structure, desired for subsequent heat treating or machining.

Hardening. Heat to 1475 °F (800 °C). Quench in water or brine. Oil quench sections under 3/16 in. (1.58 mm) for hardening.

Tempering. As-quenched hardness as high as 66 HRC. Can be adjusted downward by tempering as shown by the curve in Fig. 4.

Austempering. Responds well to austempering (bainitic hardening). Austenitize at 1475 °F (800 °C). Quench in agitated molten salt bath at 600 °F (315 °C). Hold for 2 h. Cool in air.

12L14 — Recommended Heat Treating Practice

This grade represents one of the most free-machining grades that has ever been developed—resulfurized, rephosphorized, and leaded. This grade should never be considered for applications where forging or welding is involved; consequently, normalizing and annealing are seldom required.

Case hardening treatments such as carbonitriding are commonly used for parts made from 12L14 and other low-carbon free-machining grades (see steel 1008 and refer to Chapter 10).

1137 — Recommended Heat Treating Practice

Normalizing. Not usually required. If necessary, heat to 1650 °F (900 °C). Cool in air.

Annealing. Heat to 1625 °F (885 °C). Furnace cool to 1200 °F (650 °C) at a rate not exceeding 50 °F (28 °C).

Hardening. Heat to 1550 °F (845 °C). Quench in water or brine. For full hardness, oil quench sections not exceeding ⅜ in. (9.52 mm) thick.

Tempering. As-quenched hardness approximately 45 HRC. Hardness can be adjusted downward by tempering as shown in Fig. 6.

Strengthening by Cold Drawing and Stress Relieving. Grade 1137 bars and other medium-carbon resulfurized steels are frequently strengthened to desirable levels without quench hardening. Increase the draft during cold drawing by 10 to 35% above normal. Stress relieve by heating at approximately 600 °F (315 °C). Produces yield strengths of up to 100 ksi (690 MPa) or higher in bars up to about ¾ in. (19.05 mm) diam. Machinability is very good.

1141 — Recommended Heat Treating Practice

Normalizing. Not usually required. If necessary, heat to 1625 °F (885 °C). Cool in air.

Annealing. Usually purchased by fabricator in condition for machining. If required, may be annealed by heating to 1550 °F (845 °C). Cool in furnace to 1200 °F (650 °C) or lower at a rate not exceeding 50 °F (28 °C) per hour.

Hardening. Austenitize at 1525 °F (830 °C). Quench in water or brine. For full hardness, oil quench sections less than ⅜ in. (9.52 mm) thick.

Tempering. Depending on precise carbon content, as-quenched hardness usually 48 to 52 HRC. Hardness can be reduced by tempering (see curve in Fig. 6).

Grade 1141 is also amenable to strengthening by cold drawing and stress relieving as described above for 1137.

1144 — Recommended Heat Treating Practice

This grade is generally considered a special-purpose steel. As shown in Table 5, it has a very high sulfur content, equal to free-machining 1213, thus permitting extremely fast machining with heavy cuts. Machined finishes are unusually good. High sulfur content reduces transverse impact and ductility. Can be drawn by heavy drafts at elevated temperature, 700 °F (370 °C), which results in relatively high strength and high hardness, up to 35 HRC. Machinability is excellent. Widely used for producing machined parts put in service without heat treatment. Can be purchased in the cold drawn condition and heat treated. Depending on precise carbon content, as-quenched and fully hardened 1144 should be approximately 52 to 55 HRC. Never used where forging or welding is involved.

Normalizing. Seldom required. If necessary, heat to 1625 °F (885 °C). Cool in air.

Annealing. Seldom necessary. May be annealed by heating to 1550 °F (845 °C). Furnace cool to 1200 °F (650 °C) at a rate not exceeding 50 °F (28 °C) per hour.

Hardening. Austenitize at 1525 °F (830 °C). Quench in water or brine. For full hardness, oil quench sections less than ⅜ in. (9.52 mm) thick.

Tempering. Depending on precise carbon content, as-quenched hardness usually 52 to 55 HRC. Hardness can be reduced by tempering, as shown in Fig. 6.

1151 — Recommended Heat Treating Practice

Normalizing. Seldom used. If necessary, heat to 1600 °F (900 °C). Cool in air.

Annealing. Seldom used. If necessary, heat to 1550 °F (870 °C). Cool in furnace at a rate not exceeding 50 °F (28 °C) per hour to 1200 °F (650 °C).

Hardening. Austenitize at 1525 °F (830 °C). Quench in water or brine. Oil

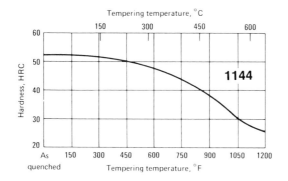

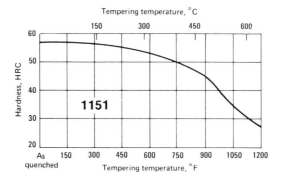

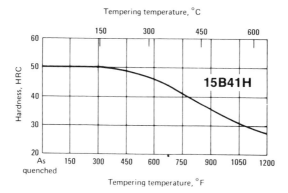

Fig. 6 Tempering curves for several of the higher carbon grades of the 11XX and 15XX series. Source: *Heat Treater's Guide: Standard Practices and Procedures for Steel*, American Society for Metals, p 87, 89, 90, 91, 106, 110, 112, 1982.

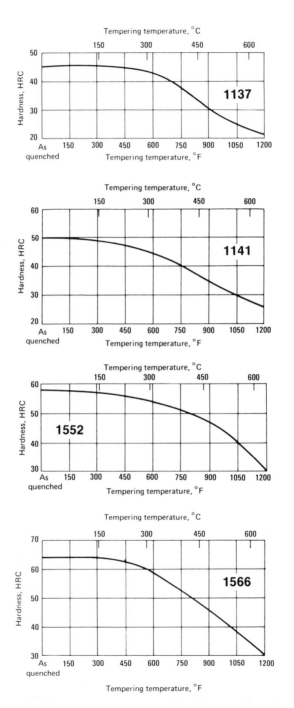

Fig. 6 (continued) Tempering curves for several of the higher carbon grades of the 11XX and 15XX series.

quench sections under ¼ in. (6.35 mm) thick.

Tempering After Hardening. Hardness of at least 55 HRC, if properly austenitized and quenched. Hardness can be adjusted by tempering, as shown in Fig. 6.

1522, 1522H — Recommended Heat Treating Practice

Normalizing. Heat to 1700 °F (925 °C). Air cool.

Annealing. Heat to 1600 °F (870 °C). Cool slowly, preferably in the furnace.

Hardening. Can be case hardened by any one of several processes, from light case hardening by carbonitriding and salt bath nitriding described for grade 1008 to deeper case carburizing in gas, solid, or liquid media. See carburizing process described for grade 1020. For 1522H, use carburizing temperature of 1650 to 1700 °F (900 to 925 °C). Use oil for cooling medium. Because of higher hardenability when compared to 1020, thicker sections can be oil quenched for full hardness. Refer to case hardening procedure given for grade 1020 and also Chapter 10.

Tempering. Temper at 300 °F (150 °C), or higher for 1 h, if some sacrifice of hardness can be tolerated.

15B41H — Recommended Heat Treating Practice

Normalizing. Heat to 1650 °F (900 °C) and cool in air.

Annealing. Heat to 1525 °F (830 °C). Furnace cool at a rate not exceeding 50 °F (28 °C) per hour to 1200 °F (650 °C).

Hardening. Heat to 1550 °F (845 °C). Should be regarded as an alloy steel in quenching from austenitizing temperature, because of the boron addition, which greatly increases hardenability compared with conventional 1541 steel (see Fig. 1). Therefore, oil quenching is usually practiced for parts made from 15B41.

Tempering After Hardening. As-quenched hardness is generally around 52 HRC, which can be reduced as desired by tempering (see Fig. 6).

Tempering After Normalizing. For large sections, normalize by conventional practice. Results in structure of fine pearlite. Temper up to about 1000 °F (540 °C). Mechanical properties attained by this treatment are not equal to those obtained by quenching and tempering, but are much better than the coarse pearlitic structure that results from annealing.

1552 — Recommended Heat Treating Practice

Normalizing. Heat to 1650 °F (900 °C). Cool in air.

Annealing. Heat to 1525 °F (830 °C). Cool in furnace at a rate not exceeding 50 °F (28 °C) per hour to 1200 °F (650 °C).

Hardening. Heat to 1525 °F (830 °C). Because of higher hardenability provided by the high manganese content, oil quench heavier sections for full hardness. When induction hardening, use the least severe quench that will produce full hardness. This minimizes the possibility of quench cracking.

Tempering. As-quenched hardness of 58 to 60 HRC can be reduced by tempering, as shown in Fig. 6.

1566 — Recommended Heat Treating Practice

This grade is simply a slightly higher manganese version of 1070. Therefore, it is used for purposes that parallel the application of 1070, namely springs and a large variety of hand tools including saws and hammers, or similar applications requiring more hardenability than can be provided by 1070.

Normalizing. Heat to 1625 °F (885 °C). Cool in air.

Annealing. Heat to 1525 °F (830 °C). Furnace cool to 1200 °F (650 °C) at a rate not exceeding 50 °F (28 °C) per hour.

Hardening. Heat to 1500 °F (815 °C). Lessen severity of quench to avoid cracking compared with quenching of 1070.

Tempering. Maximum hardness near 65 HRC can be reduced by tempering, as shown in Fig. 6.

Austempering. Thin sections (typically springs) are commonly austempered, resulting in bainitic structure and a hardness range of approximately 46 to 52 HRC. Austenitize at 1500 °F (815 °C). Quench in molten salt bath at 600 °F (315 °C). Hold at temperature for 1 h. Air cool. No tempering is required.

Martensitic and Nonmartensitic Structures

Throughout this chapter, as well as in the preceding chapters, it has been stressed that the ideal condition is to attain a 100% martensitic structure from the quench. This may be an objective of quenching, but when one examines the TTT curve presented in Chapter 2 and a similar curve in Fig. 5, it is apparent that 100% martensite is impossible to obtain in carbon steels, except in very thin sections. To do so, quenching must be severe to exceed rapid critical cooling rates.

Therefore, in practice, the objective of forming 100% martensite in the as-quenched parts is not achieved except in very thin sections. As section thickness increases, the resulting microstructure becomes more of a mixture. However, even in quenching very heavy sections, formation of coarse pearlite can be avoided (near the top of Fig. 5), thus forming fine pearlite, probably some bainite and perhaps some martensite, depending on the pattern of the cooling curve attained. Note that the hardness of the quenched product gradually decreases (right side of Fig. 5) from martensite to coarse pearlite; that is, the lower the temperature where transformation takes place, the harder the quenched product.

Some massive parts, such as large shaft forgings made from steels similar to 1045 or 1050, are not quenched at all, but are simply normalized by heating to about 1650 °F (900 °C) and by cooling in still air. They are then reheated (tempered) to approximately 1000 °F (540 °C).

As a rule, such treatments result in a structure of fine pearlite, which has good mechanical properties even though martensite is not formed.

Tempering of Quenched Carbon Steels

All parts made from carbon steels (or from quench-hardened steels) should be tempered immediately after cooling to near room temperature by quenching. This prevails for through-hardened, as well as case hardened, parts (carburized or carbonitrided specifically). It is, however, far more important that through-hardened parts are tempered; this becomes increasingly so as the carbon content increases. Many carburized or carbonitrided parts are never tempered, but this is not considered good practice.

The need for tempering also increases as the amount of martensite formed increases. Martensite is formed when carbon atoms are trapped in the iron lattice. When a high-carbon martensite is formed, the iron lattice actually has a tetragonal shape (instead of strictly cubic). In this condition, it is metastable and extremely brittle (fragile). The structure is known as white martensite because of the way it responds to etchants when prepared for microscopic examination. This metastable martensite changes to some degree, even at room temperature in a matter of time—the old timers called it "seasoning." The change from tetragonal to cubic martensite can be accomplished quickly, however, by heating. At approximately 185 °F (85 °C), the change begins, generally known as the first stage of tempering. This heating procedure brings about a marked decrease in brittleness without decreasing hardness as measured by indentation. Before heat treating developed from an art to a science, toolmakers heated and quenched tools in boiling water, making them far less brittle.

Although 185 °F (85 °C) does represent the beginning of tempering, a minimum tempering temperature of at least 300 °F (150 °C) is nearly standard because hardness is not lost from this tempering temperature (Fig. 4 and 6).

Curves that show the relationship of tempering temperature and hardness for 14 different carbon steels are presented in Fig. 4 and 6. All hardness readings were taken at room temperature after cooling from the tempering operation. Also, all data are based on thin sections wherein 100% (or near 100%) martensite is present. Depending also on the precise carbon contents within the allowable ranges, the as-quenched hardness may be slightly higher or lower than those shown. As the amount of nonmartensitic products increases, the as-quenched hardness is lower, and the entire curve drops slightly.

In examining Figs. 4 and 6, note that the hardness for any given grade drops gradually, which is characteristic for all carbon steels. This condition results from a gradual decomposition of the martensite as tempering temperature increases.

Selection of tempering temperature depends entirely on the mechanical requirements, but there are two simple rules that should always be observed in establishing a tempering temperature. A tempering temperature should be used that is as high as can be tolerated, considering the resulting hardness, and the need for a specific hardness in service. A tempering temperature should always be used that is at least as high as, and preferably slightly higher, than the maximum service temperature to which the parts will be exposed in service.

Austempering of Steel (Ref 10)

Austempering is the isothermal transformation of a ferrous alloy at a temperature below that of pearlite formation and above that of martensite formation (see Fig. 5). Steel is austempered by being:

- Heated to a temperature within the austenitizing range, usually 1450 to 1600 °F (790 to 870 °C)
- Quenched in a bath maintained at a constant temperature, usually in the range of 500 to 750 °F (260 to 400 °C)
- Allowed to transform isothermally to bainite in this bath
- Cooled to room temperature, usually in still air

The fundamental difference between austempering and conventional quenching and tempering is shown schematically in Fig. 7.

A conventional heating, quenching, and tempering cycle is shown at the left. Following the black lines down from Ae_3, note that the lines miss the nose of the TTT curve, continue straight through the martensite formation zone (M_s to M_f), then remain at constant temperature until transformation is complete, then reheating for tempering, holding time and cooling from the tempering temperature. Thus, the left side of Fig. 7 represents a conventional austenitizing, quenching, and tempering cycle, wherein 100% martensite would be formed.

On the right side of Fig. 7, the workpiece is quenched past the nose of the TTT curve, but cooling is arrested at a temperature just above the martensite zone, the workpiece is then allowed to transform isothermally after which it is cooled in air. Under these conditions, the transformation is to bainite rather than martensite (see Fig. 5). Tempering is not required.

As a rule, parts that have been austempered have greater toughness at the same hardness level compared with quenched and tempered counterparts. Other advantages of the austempering process are:

- Reduced distortion, which lessens subsequent machining time, stock removal and cost
- The shortest overall time cycle to through harden within the hardness range of 35 to 55 HRC, with resulting savings in energy and capital investment

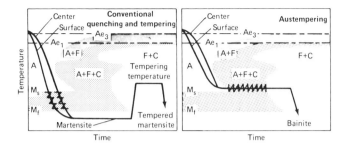

Fig. 7 Schematic comparison of time-temperature transformation cycles for conventional quenching and tempering and for austempering. Source: *Metals Handbook*, Vol 4, 9th edition, American Society for Metals, p 104, 1980.

Quenching Mediums. Molten salt is the quenching medium most commonly used in austempering because (1) it transfers heat rapidly; (2) it virtually eliminates the problem of a vapor phase barrier during the initial stage of quenching; (3) its viscosity is uniform over a wide range of temperature; (4) its viscosity is low at austempering temperatures (near that of water at room temperature), thus minimizing dragout losses; (5) it remains stable at operating temperatures and is completely soluble in water, thus facilitating subsequent cleaning operations; and (6) the salt can be easily recovered from wash waters so that there is no discharge to drain.

Formulations and characteristics of two typical salt quenching baths are given in Table 7. The high-range salt is suitable for austempering only, whereas the wide-range salt may be used for austempering.

Selection of Grade for Austempering. The selection of steel for austempering must be based on transformation characteristics as indicated in time-temperature transformation (TTT) diagrams. Three important considerations are (1) the location of the nose of the TTT curve and the time available for bypassing it, (2) the time required for complete transformation of austenite to bainite at the austempering temperature, and (3) the location of the M_s point.

As indicated in Fig. 5, 1080 carbon steel possesses transformation characteristics that provide it with limited suitability for austempering. Cooling from the austenitizing temperature to the austempering bath must be accomplished in about 1 s to avoid the nose of the TTT curve, and thus prevent transformation to pearlite during cooling. Depending on the temperature, isothermal transformation in the bath is completed within a period ranging from a few minutes to about 1 h. Because of the rapid cooling rate required, austempering of 1080 can be successfully applied only to thin sections of about 5 mm (0.2 in.) maximum.

Alloy steels, because of their greater hardenability, are often better suited for austempering than are carbon steels (see Chapter 6).

Table 7 Compositions and characteristics of salts used for austempering

Source: Metals Handbook, Vol 4, 9th edition, American Society for Metals, p 105, 1980.

	High range	Wide range
Sodium nitrate	45 to 55%	15 to 25%
Potassium nitrate.......................	45 to 55%	45 to 55%
Sodium nitrite..........................	· · ·	25 to 35%
Melting point (approx)	220 °C (430 °F)	150 to 165 °C (300 to 330 °F)
Working temperature range...............	260 to 595 °C (500 to 1100 °F)	175 to 540 °C (345 to 1000 °F)

Applications of Austempering. This process is substituted for conventional quenching and tempering for either or both of two reasons: (1) to obtain improved mechanical properties (particularly higher ductility or notch toughness at a given high hardness), and (2) to decrease the likelihood of cracking and distortion. In some applications, austempering is less expensive than conventional quenching and tempering. This is most likely when small parts are treated in an automated setup wherein conventional quenching and tempering comprise a three-step operation—that is, austenitizing, quenching and tempering. Austempering requires only two processing steps: austenitizing and isothermal transformation in an austempering bath.

The range of austempering applications generally encompasses parts fabricated from bars of small diameter or from sheet or strip of small cross section. Austempering is particularly applicable to thin-section carbon steel parts requiring exceptional toughness at a hardness near 50 HRC.

Chapter 6

Heat Treating
of Alloy Steels

It is logical for one to assume that alloy steels are steels that contain alloying elements other than carbon. While this assumption is technically correct, it is not consistent with the nomenclature generally used by the metalworking community.

Stainless steels, a majority of the tool steels, and many special-purpose steels are highly alloyed, but generally are designated for use in specific application, such as heat-resistant grades.

In general, and for purposes of this discussion, alloy steels refer to those steel grades for which specific alloy ranges have been established by American Iron and Steel Institute-Society of Automotive Engineers (AISI-SAE).

What Are Alloy Steels?

A steel is considered to be an alloy steel when the maximum of the range given for the content of alloying elements exceeds one or more of the following limits:

> Manganese 1.65%
> Silicon 0.60%
> Copper 0.60%

A steel is also classified as an alloy steel when a definite range or a definite minimum quantity of any of the following elements is specified or required within recognized limits:

- Aluminum
- Boron
- Chromium (up to 3.99%)
- Cobalt
- Niobium

- Molybdenum
- Nickel
- Titanium
- Tungsten
- Vanadium
- Zirconium

In addition to these elements, any other element added to obtain a desired effect is characterized as an alloying element.

Designation. Alloy steels also are identified by the AISI coding system, using the same general procedure described for carbon and resulfurized steels. The first two digits of the four-numeral series for the various grades of alloy steel and their meaning are given in Table 1.

The last two digits of the four-numeral series are intended to indicate the approximate middle of the carbon range. It is necessary, however, to deviate from this rule and to interpolate numbers in the case of some carbon ranges and for variations in manganese, sulfur, chromium, or other elements.

The prefix letter E is used to designate steels made by the basic electric furnace process. Steels without the prefix letter normally are manufactured by the basic open hearth or basic oxygen process. When boron is specified, the letter B is inserted after the first two numbers; for example, 41B30. When lead is specified, the letter L is inserted after the first two numbers; for example, 41L30.

Many alloy steels are specialty steels uniquely suited for certain applications. For instance, 52100 is used almost exclusively for antifriction bearings, and the 9200 series alloys are used for springs and applications where shock resistance is a factor. In most instances, however, a single composition of steel serves a wide range of applications; for example, the 4340, 8640, and 8740 grades.

H-Steels. Many of the standard grades of alloy steels are guaranteed to meet the requirements of the AISI hardenability bands. Such steels carry the suffix letter H; for example, 4140H. Therefore, if a steel must meet standard hardenability requirements, it should be so specified. However, in most

Table 1 Numerical designations of AISI and SAE grades of alloy steels
Source: Heat Treater's Guide: Standard Practices for Steel, American Society for Metals, p 15, 1982.

Series designation	Type and approximate percentages of identifying elements
13XX	Manganese, 1.75
40XX	Molybdenum, 0.20 or 0.25
41XX	Chromium, 0.50, 0.80 or 0.95; molybdenum 0.12, 0.20 or 0.30
43XX	Nickel 1.83, chromium 0.50 or 0.80, molybdenum 0.25
44XX	Molybdenum 0.53
46XX	Nickel 0.85 or 1.83, molybdenum 0.20 or 0.25
47XX	Nickel 1.05, chromium 0.45, molybdenum 0.20 or 0.35
48XX	Nickel 3.50, molybdenum 0.25
50XX	Chromium 0.40
51XX	Chromium 0.80, 0.88, 0.93, 0.95 or 1.00
5XXXX	Carbon 1.04, chromium 1.03 or 1.45
61XX	Chromium 0.60 or 0.95, vanadium 0.13 or 0.15 min.
86XX	Nickel 0.55, chromium 0.50, molybdenum 0.20
87XX	Nickel 0.55, chromium 0.50, molybdenum 0.25
88XX	Nickel 0.55, chromium 0.50, molybdenum 0.35
92XX	Silicon 2.00
50BXX	Chromium 0.25 or 0.50
51BXX	Chromium 0.80
81BXX	Nickel 0.30, chromium 0.45, molybdenum 0.12
94BXX	Nickel 0.45, chromium 0.40, molybdenum 0.12

Note: B denotes boron steel.

instances, the standard grades meet, or at least come close to meeting, the hardenability limits established by AISI.

AISI-SAE Compositions. Approximately 64 standard compositions of alloy steels are listed by AISI-SAE. In many instances, there are only minor variations among grades. For this reason, all of them are not listed in this chapter. Approximately one half (which is representative) of the standard alloy steel compositions are listed in Table 2. For a complete list, see Ref 7. Compositions of standard boron alloy steels are listed in Table 3.

Free-Machining Alloy Steels. As a rule, alloy steels are used instead of carbon steels to improve certain mechanical properties, either as produced or through heat treatment. Therefore, for most applications, users hesitate to specify alloy steels that contain free-machining additives. At best, these additives lower steel quality to some extent, thus partially defeating the purpose of using alloy steels. There are, however, notable exceptions. Consequently, most alloy steels are available with additions of lead or sulfur by special order.

Purposes Served by Alloying Elements

As a rule, the alloys contained in AISI-SAE alloy steels include manganese and silicon (over specified amounts), nickel, chromium, molybdenum, and vanadium, but in a variety of combinations. Boron, because it is used in such small quantities, is not considered an alloy. Other alloying elements such as copper, cobalt, tungsten, and titanium are not usually specified in alloy steels, but are used in stainless and tool steels (see Chapters 7 and 8).

Although there may be several reasons for using certain alloy steels as opposed to carbon steels of the same carbon content, hardenability is by far the most common reason. Hardenability controls the mechanical properties that can be obtained by heat treatment.

Although a few alloy steels contain sufficient quantities of alloying elements to provide some resistance to heat and/or corrosion compared with their carbon steel counterparts, they are not usually selected because of these properties. Only a few of the alloy grades contain a total alloy content of near 5%. For the most part, their total alloy content is substantially less than 5%.

Effects of Specific Elements

In many instances, the effects of small amounts of two or more alloying elements used together are greater than larger amounts of a single element (most notably on hardenability).

Manganese is an important alloy in steel for several reasons and is present in virtually all steels in amounts of 0.30% or more. Manganese is a carbide former and has a marked effect on slowing the gamma-to-alpha transformation; therefore, it increases hardenability. Further, manganese in steel is important because of its ability to counter hot shortness; that is, the

Table 2 Compositions of standard alloy steels

Source: Heat Treater's Guide: Standard Practices and Procedures for Steel, p 18, 1982.

Steel designation AISI or SAE	UNS No.	Chemical composition, % C	Mn	P max	S max	Si	Ni	Cr	Mo
1330	G13300	0.28-0.33	1.60-1.90	0.035	0.040	0.15-0.30	...	...	...
1345	G13450	0.43-0.48	1.60-1.90	0.035	0.040	0.15-0.30	...	...	...
4023	G40230	0.20-0.25	0.70-0.90	0.035	0.040	0.15-0.30	...	...	0.20-0.30
4027	G40270	0.25-0.30	0.70-0.90	0.035	0.040	0.15-0.30	...	...	0.20-0.30
4037	G40370	0.35-0.40	0.70-0.90	0.035	0.040	0.15-0.30	...	...	0.20-0.30
4047	G40470	0.45-0.50	0.70-0.90	0.035	0.040	0.15-0.30	...	...	0.20-0.30
4118	G41180	0.18-0.23	0.70-0.90	0.035	0.040	0.15-0.30	...	0.40-0.60	0.08-0.15
4130	G41300	0.28-0.33	0.40-0.60	0.035	0.040	0.15-0.30	...	0.80-1.10	0.15-0.25
4140	G41400	0.38-0.43	0.75-1.00	0.035	0.040	0.15-0.30	...	0.80-1.10	0.15-0.25
4145	G41450	0.43-0.48	0.75-1.00	0.035	0.040	0.15-0.30	...	0.80-1.10	0.15-0.25
4150	G41500	0.48-0.53	0.75-1.00	0.035	0.040	0.15-0.30	...	0.80-1.10	0.15-0.25
4161	G41610	0.56-0.64	0.75-1.00	0.035	0.040	0.15-0.30	...	0.70-0.90	0.25-0.35
4320	G43200	0.17-0.22	0.45-0.65	0.035	0.040	0.15-0.30	1.65-2.00	0.40-0.60	0.20-0.30
4340	G43400	0.38-0.43	0.60-0.80	0.035	0.040	0.15-0.30	1.65-2.00	0.70-0.90	0.20-0.30
4615	G46150	0.13-0.18	0.45-0.65	0.035	0.040	0.15-0.30	1.65-2.00	...	0.20-0.30
4620	G46200	0.17-0.22	0.45-0.65	0.035	0.040	0.15-0.30	1.65-2.00	...	0.20-0.30
4815	G48150	0.13-0.18	0.40-0.60	0.035	0.040	0.15-0.30	3.25-3.75	...	0.20-0.30
4820	G48200	0.18-0.23	0.50-0.70	0.035	0.040	0.15-0.30	3.25-3.75	...	0.20-0.30
5120	G51200	0.17-0.22	0.70-0.90	0.035	0.040	0.15-0.30	...	0.70-0.90	...
5140	G51400	0.38-0.43	0.70-0.90	0.035	0.040	0.15-0.30	...	0.70-0.90	...
5150	G51500	0.48-0.53	0.70-0.90	0.035	0.040	0.15-0.30	...	0.70-0.90	...
5160	G51600	0.56-0.64	0.75-1.00	0.035	0.040	0.15-0.30	...	0.70-0.90	...
E51100	G51986	0.98-1.10	0.25-0.45	0.025	0.025	0.15-0.30	...	0.90-1.15	...
E52100	G52986	0.98-1.10	0.25-0.45	0.025	0.025	0.15-0.30	...	1.30-1.60	...
6118	G61180	0.16-0.21	0.50-0.70	0.035	0.040	0.15-0.30	...	0.50-0.70	0.10-0.15 V
6150	G61500	0.48-0.53	0.70-0.90	0.035	0.040	0.15-0.30	...	0.80-1.10	0.15 V min
8620	G86200	0.18-0.23	0.70-0.90	0.035	0.040	0.15-0.30	0.40-0.70	0.40-0.60	0.15-0.25
8627	G86270	0.25-0.30	0.70-0.90	0.035	0.040	0.15-0.30	0.40-0.70	0.40-0.60	0.15-0.25
8630	G86300	0.28-0.33	0.70-0.90	0.035	0.040	0.15-0.30	0.40-0.70	0.40-0.60	0.15-0.25
8640	G86400	0.38-0.43	0.75-1.00	0.035	0.040	0.15-0.30	0.40-0.70	0.40-0.60	0.15-0.25
8655	G86550	0.51-0.59	0.75-1.00	0.035	0.040	0.15-0.30	0.40-0.70	0.40-0.60	0.15-0.25
8720	G87200	0.18-0.23	0.70-0.90	0.035	0.040	0.15-0.30	0.40-0.70	0.40-0.60	0.20-0.30
8740	G87400	0.38-0.43	0.75-1.00	0.035	0.040	0.15-0.30	0.40-0.70	0.40-0.60	0.20-0.30
8822	G88220	0.20-0.25	0.75-1.00	0.035	0.040	0.15-0.30	0.40-0.70	0.40-0.60	0.30-0.40
9260	G92600	0.56-0.64	0.75-1.00	0.035	0.040	1.80-2.20	...	...	...

tendency to tear when being hot formed. In steel, iron and sulfur combine to form a sulfide that has a relatively low melting point. When the steel freezes, this sulfide solidifies in the grain boundaries. On subsequent reheating for rolling or forging, the sulfide melts. The steel is then weakened, and when it is hot worked by rolling or forging, it tears or breaks apart. If manganese is added, it combines preferentially with the sulfur and forms a manganese sulfide of higher melting point, which by its distribution and nature eliminates hot shortness.

Silicon, to a very mild extent, retards critical cooling rate, thereby increasing hardenability. Silicon is used as a deoxidizer in steelmaking and therefore exists in small quantities—usually 0.15 to 0.30% as shown in Table 2. Silicon, as an alloy, is not extensively used in alloy steels although it has some strengthening effect. Silicon structural steels have seen considerable use. Silicon also improves shock resistance and is used where impact is a problem. It also is used as an alloy in spring steels (note grade 9260 in Table 2).

Nickel has a marked effect on the transformation of austenite by shifting the nose of the TTT curves to the right, thus increasing hardenability. It also lowers the gamma-to-alpha transformation temperature to the point that steel may become austenitic at room temperature when large amounts of nickel are used.

Nickel is a very versatile alloy. It provides an added degree of uniformity to quenched steels and strengthens the unquenched or annealed steels. It toughens ferritic-pearlitic steels, especially at low temperature, and gives good fatigue resistance. Nickel greatly increases corrosion and oxidation resistance. The amount of nickel used in alloy steels is usually less than 1.0%; the 43XX and 48XX grades are exceptions.

Chromium affects the properties of steel in a number of ways. It has a marked effect in slowing the rate of transformation of austenite; that is, it greatly increases the hardenability of any steel. Additionally, relatively large percentages of chromium greatly increase oxidation and corrosion resistance. However, because the amount of chromium used in alloy steels is 2.0% or less, its principal function for the AISI-SAE alloy steels is to increase hardenability.

Molybdenum, like chromium, greatly slows the gamma-to-alpha transformation; thus, it also increases hardenability. Molybdenum is, however, far more potent than chromium as indicated by the small amounts (generally less than 0.40%) used. Most benefit of molybdenum is achieved when used with nickel and/or chromium.

Vanadium is a strong deoxidizing agent that promotes fine grain size. In larger amounts (see Chapter 8), vanadium is a powerful carbide former that slows the gamma-to-alpha transformation, thus increasing hardenability. The amount of vanadium used in alloy steels is not, however, sufficient to form carbide. Consequently, it functions only as a grain refiner. As may be noted from Table 3, vanadium is specified only in the 61XX steels and even then in very small amounts.

Effects of Alloys on Hardenability

While alloying elements are sometimes used alone, they frequently are used in combinations of two or more elements, because it has been found that the use of smaller amounts of two or more elements is more effective than large amounts of a single element. For instance, chromium is generally used with nickel, molybdenum, or some other element. Regardless of the type or amount of the alloy added, the general principles of austenite formation and transformation remain the same, although the reference points, such as the critical temperatures and the time and temperature of transformation, vary with the type and the amount of alloying element.

The main reason for using an alloy steel as opposed to its carbon steel counterpart is to obtain greater hardenability. There are many cases where alloy steels are really wasted; that is, they are used because a poorly informed user assumes that alloy steels are "better steels" than carbon steels. This is not necessarily true. One steel is "better" than another only because it is better suited to a specific application. In the annealed condition, all other factors being equal, mechanical properties of alloy steels are only slightly higher than carbon steels with the same carbon content. This condition prevails for heat treated parts that have very thin sections. However, as section thickness increases, the necessity for more hardenability becomes more important.

Table 3 Compositions of standard boron (alloy) steels

Source: Heat Treater's Guide: Standard Practices and Procedures for Steel, p 19, 1982.

Steel designation AISI or SAE	UNS No.	C	Mn	P max	S max	Si	Ni	Cr	Mo
50B44	G50441	0.43-0.48	0.75-1.00	0.035	0.040	0.15-0.30	...	0.40-0.60	...
50B46	G50461	0.44-0.49	0.75-1.00	0.035	0.040	0.15-0.30	...	0.20-0.35	...
50B50	G50501	0.48-0.53	0.75-1.00	0.035	0.040	0.15-0.30	...	0.40-0.60	...
50B60	G50601	0.56-0.64	0.75-1.00	0.035	0.040	0.15-0.30	...	0.40-0.60	...
51B60	G51601	0.56-0.64	0.75-1.00	0.035	0.040	0.15-0.30	...	0.75-0.90	...
81B45	G81451	0.43-0.48	0.75-1.00	0.035	0.040	0.15-0.30	0.20-0.40	0.35-0.55	0.08-0.15
94B17	G94171	0.15-0.20	0.75-1.00	0.035	0.040	0.15-0.30	0.30-0.60	0.30-0.50	0.08-0.15
94B30	G94301	0.28-0.33	0.75-1.00	0.035	0.040	0.15-0.30	0.30-0.60	0.30-0.50	0.08-0.15

Relative Hardenability for Alloy Grades

As indicated from the wide variation in alloy content of the steels listed in Table 2, an equally wide variation in hardenability exists among the alloy grades. The alloy grades with the lowest hardenability are generally the 40XX series. For example, the hardenability of 4037H (shown in Fig. 1a) is little more, if any, than 1038H carbon steel.

Hardenability bands for three widely used alloy steels are presented in Fig. 1. The band for 4037H (Fig. 1a) represents the minimum hardenability for the alloy grades; the molybdenum addition of 0.20 to 0.30% is the only element that can be considered as an alloy by the established definitions.

The hardenability band for 4140H, a very popular alloy grade, is shown in Fig. 1(b). The marked effect of the chromium addition is obvious, and the even more marked effect of three alloys (nickel, chromium, and molybdenum) is shown in Fig. 1(c), the hardenability band for 4340H. This grade generally is considered the highest hardenability alloy steel. In Fig. 1(c), the upper boundary of the hardenability band is almost a straight line, which indicates that the hardenability of 4340H is approaching the hardenability of an air-hardening steel. As a practical note, a 6-in. round of 4340H steel can be austenitized and quenched in oil to produce at least 50% transformation to martensite to almost the center of the section.

The three hardenability bands in Fig. 1 clearly demonstrate how important it is to carefully consider section size in selecting an alloy grade to be heat treated.

Effect of Boron on Hardenability

While only a few standard grades of boron-treated steels are listed by AISI (see Table 3), other grades that are treated with boron commonly are available by special order.

The marked effect of boron on hardenability is best demonstrated by comparing two alloy steels that otherwise have essentially the same composition (see Fig. 2 for a comparison of the hardenability bands for 5160H and 51B60H). The use of boron-treated alloy steels offers greater hardenability without using the more highly alloyed and more expensive grades of steels.

Heat Treating Procedures

Techniques used for heat treating the alloy steels are not significantly different from those used for carbon steels described in Chapter 5. All alloy steels with carbon contents that do not exceed about 0.25% are heat treated by one of the case hardening processes. Alloy steels of higher carbon contents are austenitized by heating above the upper transformation temperature, followed by quenching to near room temperature, and finally tempering to the desired hardness level.

Major variations involved in heat treating alloy steels compared with carbon steels are:

- Temperatures used for normalizing, annealing, and austenitizing are generally at least 25 to 75 °F (12 to 42 °C) higher than for carbon steels of similar carbon content. There are some exceptions—notably the high-nickel grades like 4820.
- As alloy content increases, the annealing cycle becomes more complicated—principally, cooling from the annealing temperature must be much slower, or programmed for the specific alloy (see Ref 7) and required structure.

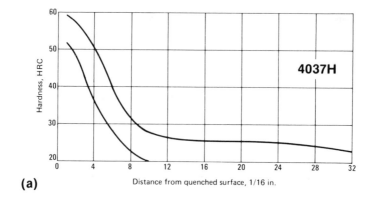

(a)

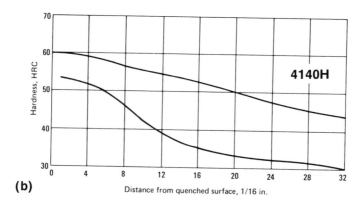

(b)

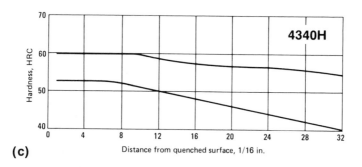

(c)

Fig. 1 Effect of total content on hardenability of three alloy steels (see text). Source: *Heat Treater's Guide: Standard Practices and Procedures for Steel,* American Society for Metals, a) p 127, b) p 146, c) p 163, 1982

- Severe quenching mediums, such as brine or water, are rarely used for alloy steels, largely because the higher hardenability of alloy steels does not require rapid cooling rates. Alloy steels are also more susceptible to cracking from severe quenching than carbon steels. Procedures for heat treating four specific, but widely different, alloy steels are described below. For more complete details and heat treating procedures for all alloy steels, see Ref 7.

4037, 4037H — Recommended Heat Treating Practice

Normalizing. Heat to 1600 °F (870 °C), and cool in air.

Annealing. For a predominantly pearlitic structure, heat to 1550 °F (845 °C), cool rapidly to 1370 °F (745 °C), then at a rate not exceeding 20 °F (11 °C) per hour to 1170 °F (630 °C); or heat to 1550 °F (845 °C), cool rapidly to 1225 °F (660 °C) and hold for 5 h. For a predominantly spheroidized structure, heat to 1400 °F (760 °C), and cool from 1370 °F (745 °C) to 1170 °F (630 °C) at a rate not exceeding 10 °F (6 °C) per hour; or heat to 1400 °F (760 °C), cool rapidly to 1225 °F (660 °C) and hold for 8 h.

Hardening. Heat to 1550 °F (845 °C), and quench in oil.

Tempering. Reheat to the temperature required to provide the desired hardness (see Fig. 3).

4140, 4140H — Recommended Heat Treating Practice

Normalizing. Heat to 1600 °F (870 °C), and cool in air.

Annealing. For a predominantly pearlitic structure, heat to 1550 °F (845 °C) and cool to 1390 °F (755 °C) at a fairly rapid rate, then cool from 1390 °F (755 °C) to 1230 °F (665 °C) at a rate not exceeding 25 °F (14 °C) per hour; or heat to 1550 °F (845 °C), cool rapidly to 1250 °F (675 °C), and hold for 5 h. For a predominantly spheroidized structure, heat to 1380 °F (750 °C) and cool to 1230 °F (665 °C) at a rate not exceeding 10 °F (6 °C) per hour; or heat to 1380 °F (750 °C), cool fairly rapidly to 1250 °F (675 °C), and hold for 9 h.

Hardening. Austenitize at 1575 °F (855 °C), and quench in oil.

Tempering. Reheat after quenching to obtain the required hardness (see Fig. 3).

4340, 4340H — Recommended Heat Treating Practice

Normalizing. Heat to 1600 °F (870 °C), and cool in air.

Annealing. For a predominantly pearlitic structure (not usually preferred for this grade), heat to 1525 °F (830 °C), cool rapidly to 1300 °F (705 °C), then cool to 1050 °F (565 °C) at a rate not exceeding 15 °F (8 °C) per hour; or heat to 1525 °F (830 °C), cool rapidly to 1200 °F (650 °C), and hold for 8 h.

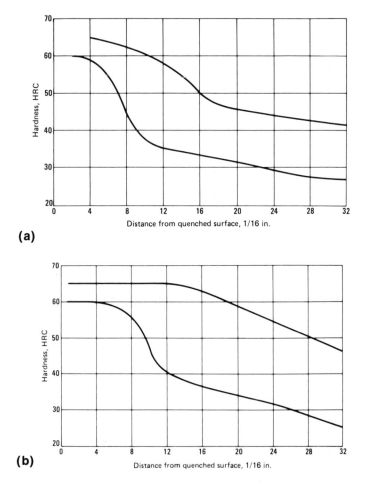

(a)

(b)

Fig. 2 Effect of boron on hardenability of 5160H alloy steel. Source: *Heat Treater's Guide: Standard Practices and Procedures for Steel,* American Society for Metals, a) p 199, b) p 202, 1982

For a predominantly spheroidized structure, heat to 1380 °F (750 °C), cool rapidly to 1300 °F (705 °C), then cool to 1050 °F (565 °C) at a rate not exceeding 5 °F (3 °C) per hour; or heat to 1380 °F (750 °C), cool rapidly to 1200 °F (650 °C) and hold for 12 h. A spheroidized structure is usually preferred for both machining and heat treating.

Hardening. Austenitize at 1550 °F (845 °C) and quench in oil. Thin sections may be fully hardened by air cooling.

Tempering. In common with all high-hardenability steels, 4340H is susceptible to quench cracking. Before parts reach ambient temperature (100 to 120 °F, 38 to 49 °C), they should be placed in the tempering furnace. Tempering temperature depends on the desired hardness or combination of mechanical properties (see Fig. 3).

E52100 — Recommended Heat Treating Practice

Normalizing. Heat to 1625 °F (885 °C) and cool in air.

Annealing. For a predominantly spheroidized structure that is generally desired for machining as well as heat treating, heat to 1460 °F (795 °C) and cool rapidly to 1380 °F (750 °C), then continue cooling to 1250 °F (675 °C) at a rate not exceeding 10 °F (6 °C) per hour; or as an alternative technique, heat to 1460 °F (795 °C), cool rapidly to 1275 °F (690 °C) and hold for 16 h.

Hardening. Austenitize at 1550 °F (845 °C) in a neutral salt bath or in a gaseous atmosphere with a carbon potential of near 1.0%, and quench in oil.

Tempering. After quenching, parts should be tempered as soon as they have uniformly reached near ambient temperature. 100 to 120 °F (38 to 49 °C) is ideal. Because of the high carbon content, parts must be tempered to at least 250 °F (120 °C) to convert the tetragonal martensite to cubic martensite. Most commercial practice calls for tempering at 300 °F (150 °C), which does not reduce the as-quenched hardness to any significant amount. When a reduction in hardness from the as-quenched value of approximately two points HRC can be tolerated, a tempering temperature of 350 °F (175 °C) is recommended. Sometimes E52100 is subjected to higher tempering temperatures, with an accompanying loss of hardness (see Fig. 3).

Effects of Tempering

From the data presented in Fig. 3, it can be seen that the as-quenched hardness of alloy steels is a function of carbon content, reaching a maximum of about 65 HRC for E52100. There is, however, some difference in the rate at which hardness decreases with increasing tempering temperature for alloy steels, as compared to carbon steels. This simply indicates the effects of alloy upon softening by increasing temperature; note especially the hardness versus tempering curve for 4340, 4340H in Fig. 3.

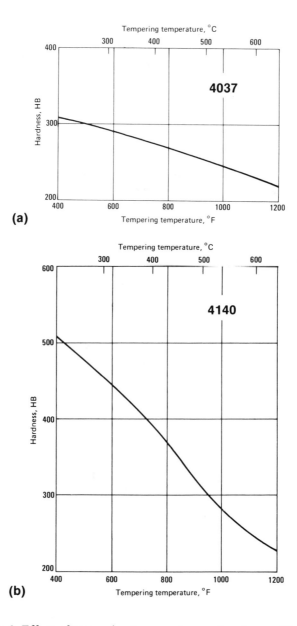

Fig. 3 Effect of tempering temperature on hardness of four alloy steels, beginning with as-quenched hardness. **4037** and **4140**: Normalized at 1600 °F (870 °C). Quenched from 1550 °F (845 °C) and tempered at 100 °F (56 °C) intervals in 0.540-in. (13.716-mm) rounds. Tested in 0.505-in. (12.827-mm) rounds. Source: *Heat Treater's Guide: Standard Practices and Procedures for Steel,* American Society for Metals, 1) 128, b) p 147, c) p 167, d) p 204, 1982

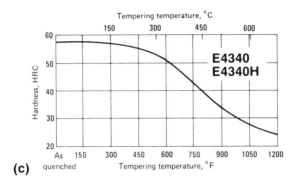

(c)

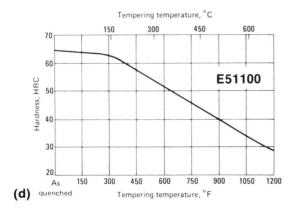

(d)

Fig. 3 (continued) **E4340**, **E4340H**, and **E51100**: Represent an average based on a fully quenched structure.

Austempering and Martempering Treatments

Many grades of alloy steels are well suited to austempering—better suited, in many instances, than carbon steels. The thicker sections of alloy steels can be successfully austempered. Alloy steels that are frequently austempered include 5150, 6150, 50B60, and 51B60. For a description of austempering, see Chapter 5.

Martempering. The term "martempering" describes an elevated-temperature quenching procedure aimed at reducing cracks, distortion, or residual stresses. It is not a tempering procedure, as the name implies, and is more properly termed "marquenching."

Martempering of steel consists of (1) quenching from the austenitizing temperature into a hot fluid medium (hot oil, molten salt, molten metal, or a fluidized particle bed) at a temperature usually above the martensitic range (M_s point); (2) holding in the quenching medium until the temperature throughout the steel is substantially uniform; and then (3) cooling (usually in air) at a moderate rate, to prevent large differences in temperature between the outside and the center of the section. Formation of martensite occurs fairly uniformly throughout the workpiece during cooling to room temperature, thereby avoiding formation of excessive amounts of residual stress. The microstructure after martempering is essentially primary martensite, which is untempered and brittle for most applications. After being air cooled to room temperature, martempered parts are tempered in the same manner as if they had been conventionally quenched.

Figure 4 shows the significant difference between conventional quenching (a) and martempering (b). The principal advantage of martempering lies in the reduced thermal gradient between surface and center as the part is quenched to the isothermal temperature and then is air cooled to room temperature. Residual stresses developed during martempering are lower than those developed during conventional quenching, because the greatest thermal variations occur while the steel is in the relatively plastic austenitic condition and because final transformation and thermal changes occur throughout the part at approximately the same time. Martempering also reduces or eliminates susceptibility to cracking.

Generally, any steel that has sufficient hardenability to harden by oil quenching can be successfully martempered. It must be emphasized that martempering does not in any way eliminate the need for tempering (see Fig. 4).

Modified martempering differs from "standard" martempering only in that the temperature of the quenching bath is below the M_s point (Fig. 4c). The lower temperature increases the severity of quenching. This is important for steels of lower hardenability that require faster cooling to harden to sufficient depth, or when the M_s is high and some bainite is detrimental to the finished part. Thus modified martempering is applicable to a greater range of steel compositions than is the standard process.

Although hot oil is invariably the quenchant employed for modified martempering at 350 °F (175 °C) and lower, molten nitrate-nitrite salts (with water addition and agitation) are effective at temperatures as low as 350 °F (175 °C). Due to their higher heat-transfer coefficients, molten salts offer some metallurgical and operational advantages. For more details on these special heat treating processes, see Ref 10.

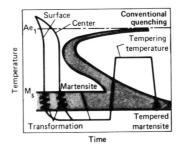

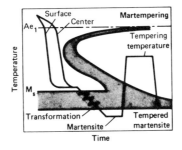

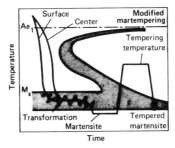

Fig. 4 Time-temperature transformation diagrams with superimposed cooling curves showing quenching and tempering. Source: *Metals Handbook*, 9th edition, Vol 4, p 86

Chapter 7

Heat Treating of Stainless Steels

The term "stainless steels" is a misnomer for two reasons: (1) none of the steels included in this grade are completely "stainless" under all conditions; and (2) only a few are truly "steels" in terms of their response to heat treatments used for the carbon and alloy steels, as discussed in Chapters 5 and 6.

Stainless steels include approximately 60 standard compositions of corrosion- and/or heat-resistant iron-base alloys, in addition to at least 100 nonstandard compositions. An iron-base alloy is one that contains 50% or more of elemental iron.

What Are Stainless Steels?

Chromium is the basis for corrosion resistance in stainless steels. One or more additional elements may be used in conjunction with chromium for most grades of stainless steels, but chromium is the key element contributing to corrosion resistance. When chromium is added to iron in relatively small amounts (1 to 3%), a modest increase in corrosion resistance of the alloy is evident. However, as the amount of chromium approaches approximately 10%, a dramatic increase in corrosion resistance takes place; such an alloy is virtually impervious to rusting from almost any outdoor exposure to normal atmospheres, such as rain, humidity, and temperature variations. This does not necessarily hold true in saline or industrial atmospheres.

Therefore, a minimum requirement for a stainless steel is that it contain at least 11.0% chromium and that it is capable of resisting attack from normal atmospheric exposure. It is essential that both of these conditions are fulfilled for a steel to qualify as a stainless steel. Many highly alloyed iron-base alloys, such as certain tool steels, contain more than 11.0% chromium, but because of their high carbon content, they do not meet the minimum requirements for stainless steels.

The mechanism of corrosion resistance in stainless steels occurs when chromium alloys with iron and forms a transparent surface oxide that serves to protect the alloy. However, formation of this protective oxide is dependent on the chromium alloying with the ferrite (iron). Carbon has a high affinity for alloying with chromium and can, as carbon content increases, "steal away" most of the chromium. This action impoverishes the ferritic matrix and impairs corrosion resistance, because chromium carbide has little resistance to most corrosive media.

Classification of Wrought Stainless Steels

The American Iron and Steel Institute (AISI) has adopted standard designation numbers for nearly 60 grades of wrought stainless steels that are divided into four groups—austenitic, ferritic, martensitic, and precipitation-hardening grades. These terms generally are based on existing structures or on structures that can be attained in the steel by heat treatment.

Compositions of several prominent grades of each of the four groups are listed in Tables 1 to 4. For more complete listings, see Ref 7. Each table also includes UNS designations.

In addition to the steels listed in Tables 1 to 4, there are at least 100 (perhaps many more) nonstandard compositions that are marketed under proprietary names. These nonstandard grades are produced in relatively small quantities; consequently, AISI does not list them as standard grades. In most instances, the nonstandard compositions have been developed to resist corrosion or heat, or both, in specific environments.

Austenitic Grades

Referring to Table 1, the austenitic grades carry identifying numbers of either 200 or 300, although most of the steels in Table 1 are 300 series alloys—the chromium-nickel grades. The 200 series represents a more recent addition to the austenitic series in which some of the nickel has been replaced by manganese. The austenitic grades are used most widely in corrosive environments, although some grades (most notably type 310) are used for elevated temperature service up to 1200 °F (650 °C).

Because the austenitic grades do not change their crystal structure on heating, they do not respond to conventional quench-hardening treatments. Therefore, the only heat treatments that are used for these grades are full annealing by rapid cooling from elevated temperatures and, in some instances, stress relieving.

The 200 and 300 grades cannot be quench-hardened like alloy steels. As shown in Table 1, all of these grades have very low carbon contents, but relatively high nickel and/or manganese contents, both of which are strong austenite (gamma) formers. Type 301 can have a combined nickel and manganese content of about 8.0% (Table 1), but for most grades shown in Table 1 the nickel and manganese (principally nickel) content is much higher (around 24% for Type 310).

Now referring to Fig. 1, the vertical axis as a line only is essentially the constitutional diagram for pure iron (see Chapter 2 for a complete explanation). Now for the moment, instead of adding carbon we add chromium in increasing amounts (horizontal axis). Now the examination must be restricted to the inner loop (vertical lines and identified on Fig. 1 as "Gamma Loop" of pure Fe-Cr alloys). Note that chromium has a marked effect on the characteristics of pure iron. As mentioned previously, quench hardening can be accomplished only when a gamma (austenite)—alpha (ferrite) transformation takes place. Now, without considering the effects of either nickel or carbon, it is evident that an alloy with approximately 12.0% chromium, when heated above about 1650 °F (900 °C) transforms to austenite, thus establishing the basis for a hardenable alloy.

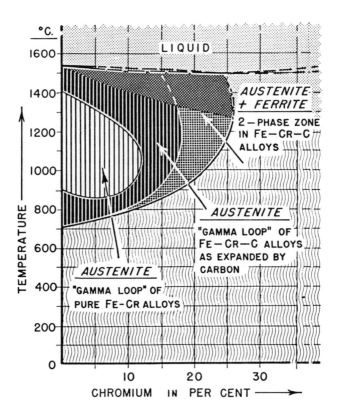

Fig. 1 Effect of carbon on the "gamma loop" of iron-chromium alloys. Source: *Stainless Steels*, American Society for Metals, p 106, 1949.

Table 1 Composition of some standard grades of wrought austenitic stainless steels

Source: Heat Treater's Guide: Standard Practices and Procedures for Steel, American Society for Metals, p 401, 1982.

Type No.	UNS No.	C	Mn	P	S	Si	Cr	Ni	Mo	Other elements
201	S20100	0.15	5.50-7.50	0.060	0.030	1.00	16.00-18.00	3.50-5.50	...	0.25 N
301	S30100	0.15	2.00	0.045	0.030	1.00	16.00-18.00	6.00-8.00	...	...
302	S30200	0.15	2.00	0.045	0.030	1.00	17.00-19.00	8.00-10.00	...	...
303	S30300	0.15	2.00	0.200	0.150 min	1.00	17.00-19.00	8.00-10.00	0.60	...
304	S30400	0.08	2.00	0.045	0.030	1.00	18.00-20.00	8.00-10.50	...	...
305	S30500	0.12	2.00	0.045	0.030	1.00	17.00-19.00	10.50-13.00	...	...
310	S31000	0.25	2.00	0.045	0.030	1.50	24.00-26.00	19.00-22.00	...	...
316	S31600	0.08	2.00	0.045	0.030	1.00	16.00-18.00	10.00-14.00	2.00-3.00	...
321	S32100	0.08	2.00	0.045	0.030	1.00	17.00-19.00	9.00-12.00	...	5 × C Ti min
347	S34700	0.08	2.00	0.045	0.030	1.00	17.00-19.00	9.00-13.00	...	10 × C Cb + Ta min

Table 2 Composition of some standard grades of wrought ferritic stainless steels

Source: Heat Treater's Guide: Standard Practices and Procedures for Steel, p 402, 1982.

Type	UNS No.	C	Mn	P	S	Si	Cr	Mo	Other elements
405	S40500	0.08	1.00	0.040	0.030	1.00	11.50-14.50	...	0.10-0.30 Al
409	S40900	0.08	1.00	0.045	0.045	1.00	10.50-11.75	...	6 × C Ti min 0.75 max
430	S43000	0.12	1.00	0.040	0.030	1.00	16.00-18.00	...	...
442	S44200	0.20	1.00	0.040	0.030	1.00	18.00-23.00	...	...
446	S44600	0.20	1.50	0.040	0.030	1.00	23.00-27.00	...	0.25 N

Table 3 Composition of some standard wrought grades of martensitic stainless steels
Source: Heat Treater's Guide: Standard Practices and Procedures for Steel, p 402, 1982.

Type	UNS No.	Chemical composition, %								Other elements
		C	Mn	P	S	Si	Cr	Ni	Mo	
410	S41000	0.15	1.00	0.040	0.030	1.00	11.50-13.50	...	...	...
414	S41400	0.15	1.00	0.040	0.030	1.00	11.50-13.50	1.25-2.50	...	...
416	S41600	0.15	1.25	0.060	0.150 min	1.00	12.00-14.00	...	0.60	...
420	S42000	Over 0.15	1.00	0.040	0.030	1.00	12.00-14.00	...	...	...
440A	S44002	0.60-0.75	1.00	0.040	0.030	1.00	16.00-18.00	...	0.75	...
440B	S44003	0.75-0.95	1.00	0.040	0.030	1.00	16.00-18.00	...	0.75	...
440C	S44004	0.95-1.20	1.00	0.040	0.030	1.00	16.00-18.00	...	0.75	...

Table 4 Composition of standard grades of precipitation-hardening stainless steels
Source: Heat Treater's Guide: Standard Practices and Procedures for Steel, p 403, 1982.

AISI No.	UNS No.	Chemical composition, %						Other elements
		C	Mn	Si	Cr	Ni	Mo	
Martensitic type								
630	S17400	0.07	1.0	1.0	17.0	4.0	...	4.0 Cu, 0.15-0.45 Cb + Ta
Semiaustenitic types								
631	S17700	0.09	1.0	1.0	17.0	7.0	...	1.0 Al
632	S15700	0.09	1.0	1.0	15.0	7.0	2.2	1.2 Al
633	S35000	0.08	0.8	0.25	16.5	4.3	2.75	0.1 N
634	S35500	0.13	0.95	0.25	15.5	4.3	2.75	0.1 N

Effects of Nickel and/or Manganese. When nickel or manganese (nickel is more frequently used) are added to an iron-chromium alloy, the "gamma loop" (Fig. 1) expands to the right and downward so that, with the amount of alloying elements used in the 200 and 300 stainless steels, the entire area down to and below room temperature in Fig. 1 is austenite. Under these conditions, no phase change takes place during heating and cooling; thus, none of these alloys can be quench hardened. They can, however, be annealed as required for fabrication, because some grades (notably Type 301) are highly susceptible to work hardening. It is now evident why the 200 and 300 series stainless steels are not really "steels" when compared with carbon and alloy steels. The procedure used for annealing austenitic stainless steels is similar to that used for hardening carbon and alloy grades; that is, austenitic stainless steels are annealed by heating to an elevated temperature, followed by rapid cooling. This results in a simple microstructure of equiaxed grains, as shown in Fig. 2.

Heat Treating of Austenitic Grades

Recommended annealing temperatures for the grades shown in Table 1 are given in Table 5. The precise heating temperature is not extremely critical, although with one exception (Type 321) the grades shown in Table 1 should be heated to at least 1850 °F (1010 °C) to ensure complete solid solution of the chromium carbides.

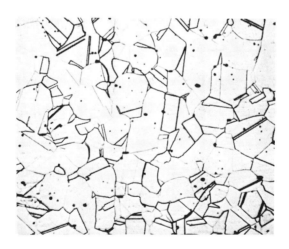

Fig. 2 Microstructure of an annealed austenitic stainless steel. HNO_3-acetic-HCl-glycerol, 250X. Strip, annealed at 1950 °F (1065 °C) for 5 min. Cooled rapidly to room temperature. Equiaxed austenite grains and annealing twins. Source: *Metals Handbook*, 8th ed., Vol 7, American Society for Metals, 1972.

Cooling practice is the critical operation in the annealing of these steels. It is essential that all of these steels be cooled from the annealing temperature to below 850 °F (450 °C) in a very short time (1 or 2 min). Allowing these steels to "linger" within the temperature range of about 850 to 1475 °F (450 to 800 °C) results in precipitation of chromium carbides at the grain boundaries (known as sensitization), which embrittles the alloy and impairs its corrosion resistance.

Exceptions to the above are types 321 and 347, which contain small additions of titanium, or niobium and tantalum, respectively. These are referred to as stabilized grades and are not susceptible to grain boundary carbide precipitation. They are used in applications where annealing is not practical, or for applications involving service temperatures within the critical temperature range stated above. Also, grades with extra-low-carbon are less susceptible to sensitization, simply because there is very little carbon to form carbides.

Cooling from the annealing temperature may be by water or oil quenching. For very thin sections, air cooling is satisfactory—the objective is to cool rapidly through the critical range. Heavier sections must be water quenched.

Stress relieving (if desired) following cold working operations can be accomplished for the austenitic grades by heating up to 750 °F (400 °C); the cooling rate from this temperature is not critical.

Corrosion Resistance of Austenitic Grades

As a group, the austenitic grades (AISI 200 and 300 series) have greater corrosion resistance than the other three groups, in terms of the number of different corrosive environments in which they remain passive (corrosion resistant). However, there are large differences in corrosion-resistant properties among the 10 grades listed in Table 1. Types 301 and 310 represent minimum and maximum corrosion resistance for the group, at least in most cases, simply because of the difference in their total chromium and nickel contents. Although chromium is the principal contributing factor to corrosion resistance, nickel is a "powerful helpmate."

Ferritic Grades

The five compositions listed in Table 2 represent the most important ferritic steels, which are identified as the 400 series. For a complete AISI listing, see Ref 7. In corrosion resistance, these steels generally rank higher than the martensitic grades, but substantially lower than most of the austenitic grades.

The ferritic grades are so named because their structure is ferritic at all temperatures, thus like the austenitic grades, they cannot be hardened by heating and quenching. Annealing to lower hardnesses developed by cold working is the only heat treatment applied to the ferritic grades.

Heat Treating of Ferritic Grades

Figure 1 also can be used to explain why the ferritic grades cannot be hardened by austenitizing and quenching. It is similar to the explanation as to why austenitic grades cannot be quench hardened; except the condition is reversed. Austenitic grades cannot be hardened because the alloy does not transform from gamma to alpha, while the ferritic grades cannot be quench hardened because there is no alpha-to-gamma transformation.

As shown in Fig. 1, if a line is drawn vertically from the horizontal at 10% chromium, it intersects the original "gamma loop" (disregard the shaded extensions of the loop). This indicates that a 10% chromium alloy can be austenitized. However, if a line is drawn vertically from the horizontal from 15% chromium, it does not intersect any portion of the "gamma loop," and therefore, it cannot be austenitized (all of the area below and to the right of the loop is ferrite, or alpha phase).

Thus far, discussion has been limited to the iron-chromium system and carbon has not been considered. As shown in Fig. 1, the expanded loop areas show the effects of increased carbon, which has a profound effect in enlarging the loop in the pattern as shown. For high carbon contents, the gamma phase (austenite), under conditions of high temperature, can still be formed with a chromium content of around 25%. Actually, the practical upper limit is usually about 20%.

Therefore, whether or not an iron-chromium-carbon alloy can be quench hardened depends greatly on the balance of carbon and chromium. This is a complex subject; consequently, a detailed discussion is not included. For more information, see Ref 14.

Type 430 (Table 2) contains a maximum of 0.12% carbon and chromium in the range of 16.0 to 18.0% — a typical ferritic steel that is widely used. Normally, this alloy cannot be hardened by quenching. However, if the carbon content is increased to 0.40%, or the chromium content is decreased, it readily becomes a quench-hardening alloy.

On a practical basis, however, when the carbon is on the high side of the allowable range and the chromium is on the low side, a borderline condition can be created whereby some austenite is formed on heating. This results in some hardening from the phase change mechanism.

Annealing of the ferritic steels can be achieved readily by the heating and cooling practice presented in Table 6. The annealing temperatures in this case are substantially lower than those used for annealing the austenitic grades. Further, the cooling rate is less critical compared with cooling rates required for annealing the austenitic grades. To avoid embrittlement, however, cooling should be as rapid as possible, as indicated by the recommended practice shown in Table 6.

Table 5 Full annealing temperatures for the more common austenitic stainless steels

Source: Heat Treater's Guide: Standard Practices and Procedures for Steel, p 406, 1982.

AISI	UNS	Annealing temperature (a) °F	°C
201	S20100	1850-2050	1010-1120
301	S30100	1850-2050	1010-1120
302	S30200	1850-2050	1010-1120
303	S30300	1850-2050	1010-1120
304	S30400	1850-2050	1010-1120
...	S30430	1850-2050	1010-1120
305	S30500	1850-2050	1010-1120
310	S31000	1900-2100	1040-1150
316	S31600	1850-2050	1010-1120
321	S32100	1750-2050	955-1120
347	S34700	1850-2050	1010-1120

(a) In all cases, cooling must be rapid, usually by water quenching, depending on section thickness.

Table 6 Annealing treatments for several ferritic stainless steels

Source: Heat Treater's Guide: Standard Practices and Procedures for Steel, p 418, 1982.

Type	Temperature(a) °F	°C	Cooling method(b)
405	1350-1500	735-815	AC or WQ
409	1600-1650	885-900	AC
430	1400-1500	760-815	AC or WQ
442	1350-1500	735-815	AC or WQ
446	1450-1600	790-870	AC or WQ

(a) Time at temperature depends on section thickness, but is usually 1 to 2 h, except for sheet which may be soaked 3 to 5 min per 0.10 in. (2.5 mm) of thickness. (b) AC, air cool; WQ, water quench

Martensitic Grades

These alloys are capable of changing their crystal structure on heating and cooling, thus responding to heat treatment much the same as carbon and alloy steels.

The martensitic group is comprised of 12 steels; compositions for 7 of these steels are listed in Table 3. Steels of this group also carry the 400 series designation. Nickel is specified in only one grade shown, and the maximum amount is 2.5%. Further, chromium content is generally lower than for the austenitic grades. Consequently, corrosion resistance of the martensitic grades is far lower compared with that of the austenitic grades and, in most instances, somewhat lower than that of the ferritic grades.

The relative corrosion resistance of the different martensitic grades is not necessarily equal, and corrosion resistance also does not necessarily increase with increasing chromium content. As indicated earlier, carbon has a high affinity for chromium, and chromium carbide contributes little if any to the total corrosion resistance of the alloy. In fact, when massive chromium carbides form (as they may in grade 440C), these carbides may be corrosion nuclei. Therefore, type 410 (nominal composition of 12.5% chromium) often has better corrosion resistance than 440C with 17.0% carbon.

Hardenability

All of the martensitic grades have extremely high hardenability, to the extent that they can be fully hardened by quenching in still air from their austenitizing temperatures, as shown in Fig. 3. Figure 3(a) is a TTT curve for type 410, which shows that the "nose" of the curve is shifted far to the right, thus indicating a very low critical cooling rate compared with carbon and alloy steels (Fig. 3b).

Hardenability is shown as a straight line; the end that was drastically quenched in water is no harder than the opposite end of the specimen which was cooled in still air (see Chapter 3).

Heat Treating of Martensitic Grades

The maximum hardness that can be achieved by austenitizing martensitic stainless steels is governed by the carbon content. The martensitic microstructures of stainless steels in the as-quenched or quenched and tempered condition are similar in appearance to those of alloy steels.

Annealing practice for the 7 grades of martensitic alloys is given in Table 7. Full and and subcritical annealing processes are shown. The subcritical approach is used whenever possible, because the cooling rate from the subcritical range is not critical. Cooling from the full annealing temperature requires a cooling rate no faster than about 30 °F (17 °C) per hour.

Austenitizing and Quenching. Austenitizing temperatures for 7 martensitic grades are given in Table 7. Temperatures in the middle of those shown usually are recommended. Air cooling frequently is used, although oil quenching may be used. For the higher carbon grades, common practice is to oil quench to "black" then air cool. This minimizes scale formation.

Especially with higher carbon grades, quenched parts should be tempered immediately, and preferably just before they reach room temperature. Ideally, when parts are "warm to the touch," the tempering operation should be started. Failure to follow this practice often results in cracking.

Tempering. All quenched parts of martensitic stainless must be tempered. These steels are highly alloyed and do not behave like carbon and alloy steels in tempering. For most alloy steels, as tempering temperature increases, the hardness decreases. This is not the case with the martensitic stainless steels.

For these high-alloy steels, transformation of the austenite is not complete upon quenching, which allows a great deal of austenite to remain as retained austenite — a metastable constituent. Some of the retained austenite transforms to martensite during tempering, which actually may result in high hardness after tempering at some elevated temperature (such as 900° F or 480

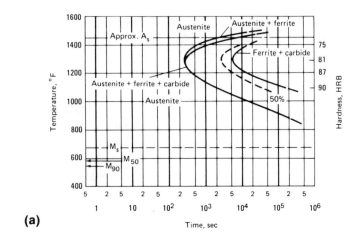

(a)

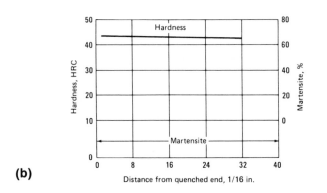

(b)

Fig. 3 Type 410 stainless steel. Composition: 0.11 C, 0.44 Mn, 0.37 Si, 0.16 Ni, 12.18 Cr. Austenitized at 1800 °F (980 °C). Grain size 6 to 7. (a) TTT curve. (b) End-quench hardenability. Source: *Atlas of Isothermal Transformation and Cooling Transformation Diagrams,* American Society for Metals, 1977.

°C). After this hardness peak, the decrease of hardness with an increase of tempering temperature is generally rapid, as shown in Fig. 4 for grade 440C. This hardness "hump," which occurs at about 900 °F (480 °C), is the result of phenomenon known as secondary hardening.

Despite the fact that higher hardnesses may be attained by hitting the peak of the secondary "hump," a decrease in corrosion resistance results when tempering temperature exceeds about 700 °F (370 °C). Therefore, 700 °F

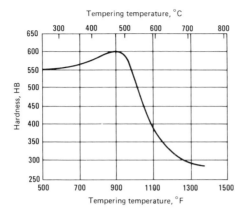

Fig. 4 Hardness versus tempering temperature for type 440C stainless steel after oil quenching from 1900 ° F (1040 ° C). Source: *Heat Treater's Guide: Standard Practices and Procedures for Steel*, American Society for Metals, p 439, 1982.

(370 °C) is usually the maximum tempering temperature — minimum 300 ° F (150 °C) as indicated in Table 7.

For all martensitic stainless steels, but more especially for the higher carbon grades, double tempering should be used. Steels should be heated to the pre-established tempering temperature, cooled to room temperature, then tempered again at the same temperature.

Precipitation-Hardening Grades

The 5 steels listed in Table 4 are the important grades of the precipitation-hardening group; they are further subdivided as to the type of structure they develop. These steels are now known as the 600 series.

In corrosion resistance, these steels may vary considerably among the different grades within the group, but generally, their corrosion resistance approaches that of some of the austenitic grades.

Most of the precipitation-hardening grades can be hardened to at least 42 HRC and higher, but not by conventional quench-hardening techniques used for martensitic grades. Hardening techniques for the precipitation-hardening grades are similar to those used for nonferrous metals. General practice is to solution treat by heating to an elevated temperature, followed by rapid cooling, then age hardening by heating to an intermediate temperature. There is, however, considerable difference in techniques used for various grades.

The metallurgy of precipitation hardening stainless steels is complex and cannot be completely covered herein. For more detailed information, see Ref

Table 7 Temperatures for heat treating several martensitic stainless steels
Source: Modified from MEI Course 18, Lesson 7, American Society for Metals, p 8, 1978.

Type		Annealing temperature		Subcritical annealing temperature		Hardening temperature		Tempering temperature	
AISI	UNS	°F	°C	°F	°C	°F	°C	°F	°C
410	S41000	1500-1650	815-900	1200-1400	650-760	1700-1850	925-1010	400-1400(a)	205-760
414	S41400	···	···	1200-1300	650-705	1800-1900	930-1040	400-1300(b)	205-705
416	S41600	1500-1650	815-900	1200-1400	650-760	1700-1850	925-1010	400-1400(a)	205-760
420	S42000	1550-1650	845-900	1350-1450	730-790	1800-1900	930-1040	300-700	150-370
440A	S44002	1550-1650	845-900	1350-1450	730-790	1850-1950	···	300-700	150-370
440B	S44003	1550-1650	845-900	1350-1450	730-790	1850-1950	···	300-700	150-370
440C	S44004	1550-1650	845-900	1350-1450	730-790	1850-1950	···	300-700	150-370

(a) Tempering in the range of 700 to 1050° F (370 to 565° C) results in decreased impact strength and corrosion resistance. (b) Tempering in the range of 750 to 1100° F (400 to 600° C) results in decreased impact strength and corrosion resistance.
Source: Modified from MEI Course 18, Lesson 7, American Society for Metals, p 8, 1978.

7 and 15. However, successful precipitation hardening depends on the presence of one or more elements that are readily soluble in the alloy at elevated temperatures, but are less soluble or insoluble at ambient temperatures.

For example, in AISI type 630, copper is the principal element responsible for precipitation hardening. Copper, in this case, is assisted by small amounts of niobium and tantalum (see Table 4).

Recommended Heat Treating Practice

Continuing with Alloy 630 as an example, largely because it is one of the most widely used of the precipitation-hardening grades, the recommended heat treating procedure is as follows. For this grade and for most other precipitation-hardening grades, the final properties differ somewhat with variations in aging temperature. However, the following procedure is most commonly used:

- Solution treat by heating at 1875 to 1925 °F (1025 to 1050 °C.
- Oil quench to room temperature.
- Age at 900 °F (480 °C) for 1 h and air cool (known as condition H900).

Following this heat treatment, a hardness of approximately 42 to 44 HRC can be expected.

Regardless of the specific alloy, it generally is as soft as it can be made following solution treating. Because the alloy is in a metastable condition, some hardening occurs even at room temperature over a long period of time. However, full hardness can be accomplished in an hour or so by heating to 900 °F (480 °C), as prescribed above.

Chapter 8

Heat Treating
of Tool Steels

It would seem logical that the term "tool steels" would refer to steels used for making tools, but this is not entirely true, depending largely on individual interpretation of what tools are and what they are not. To the layman, tools are items such as hammers, chisels, and saws commonly found in a hardware store. While these are tools, it is rare that any of these hardware store items are made from materials that conform to a manufacturer's understanding of tool steels. Tool steels represent a small, but very important, segment of the total production of steel. Their principal use is for tools and dies that are used in the manufacture of commodities.

Hundreds of nontooling applications, however, are fulfilled by use of tool steels. Most applications are components for mechanical devices that demand steels with special properties such as high resistance to abrasion or heat, or both. For example, some hot work grades of tool steels are used extensively for components for aerospace vehicles, gas turbines, and other jet-age requirements.

Classification of Tool Steels

Over 100 different compositions of tool steel are produced; the trade names for tool steels total far more than 100.

Tool steels vary in composition from simple plain carbon steels that contain iron and as much as 1.2% C with no significant amounts of alloying elements, up to and including very highly alloyed grades in which the total alloy content approaches 50%. In between these extremes, practically every combination of the principal alloying elements including manganese, silicon, chromium, nickel, molybdenum, tungsten, vanadium, and cobalt has been employed. The great diversity among tool steels has posed a major problem in classifying tool steels.

Historically, tool steels were known and marketed by trade name. Unfortunately, this practice still persists in many areas. In reviewing the compositions of tool steels, it is evident immediately that many are identical in composition to carbon and alloy steels that are produced in large tonnages. Then, why pay the higher cost for a tool steel? The only answer is to ask another question—what quality level is required? Many tooling applications can be fulfilled satisfactorily by using lower priced carbon or alloy grades; conversely, as many nontooling applications require the high level of quality provided by tool steels. It should be clearly understood that tool steels are made and processed in small quantities to extremely high levels of quality control.

Tool steels do not lend themselves to the type of classification used by SAE and AISI for carbon and alloy steels, in which an entire series of steels is defined numerically and based on a variation of carbon content alone. While some carbon tool steels and low-alloy tool steels are made in a wide range of carbon contents and permit such a classification, most of the higher alloyed types of tool steels have a comparatively narrow carbon range, and such a classification would be meaningless. Tool steels cannot be classified on the basis of the predominant alloying element, for in many of the more complex compositions, one alloying element can be partially or wholly substituted for another with little change in the mechanical properties of the steel. Classification by application appears possible for certain types of tool steels, but quite impractical for others. For example, tool steels used for hot extrusion tools can be closely classified. It would be impossible, however, to combine in one class all of the tool steels that are used to make a tap. Steels suitable for taps range from carbon tool steels through the super-high speed steels.

There are certain groups of tool steels that, while varying considerably in composition, have mechanical properties and other characteristics which are so similar that they naturally fall within a common group.

One logical approach to classification is to use a mixed classification in which some steels are grouped by use, others by composition or by certain mechanical properties, and still others by the method of heat treatment (precisely by the quenching technique). Actually, through the mediums of shop language and the published literature, tool steels have become somewhat automatically classified on such a basis. For instance, the terms high speed steels, water-hardening steels, hot work steels, and high-carbon, high-chromium steels represent just such a mixed classification.

High speed steels are grouped together because they have certain common properties; the water-hardening steels because they are hardened in a common manner; the hot work steels because they have certain common properties; and the high-carbon, high-chromium steels are grouped together because of their similar compositions and applications.

The American Iron and Steel Institute lists a total of 72 different compositions in their Steel Products Manual (Ref 16). Table 1 lists 24 of the most widely used of these compositions, which are representative of the total group. The Unified Numbering System (UNS) identifications are also given in Table 1. All elements are given in nominal amounts, which may vary somewhat for different tool steel producers. When heat treating shops receive tools for heat treatment that are identified only by proprietary name, every effort should be made to obtain the AISI identification before performing any heat treating operation.

The grouping of tool steels published by AISI has proved workable; the nine main groups and their corresponding symbols are:

Name	Identifying symbol
Water-hardening tool steels	W
Shock-resisting tool steels	S
Oil-hardening cold work tool steels	O
Air-hardening, medium-alloy cold work tool steels	A
High-carbon, high-chromium cold work tool steels	D
Mold steels	P
Hot work tool steels, chromium, tungsten, and molybdenum	H
Tungsten high speed tool steels	T
Molybdenum high speed tool steels	M

For three groups (W, O, and A), the quenching medium serves as the identifying symbol. In other instances, the use is indicated by the letter symbol. In other instances, the use is indicated by the letter symbol. In most instances, each group bears a common letter symbol as indicated above. High speed steels are the exception because of the first group, which employs tungsten as the principal alloying element, and the succeeding groups wherein molybdenum is the principal alloying element.

Heat Treating Processes for Tool Steels — General

For the most part, the processes used for heat treating carbon and alloy steels are also used for heat treating tool steels—annealing, austenitizing, tempering, etc. However, normalizing is used only to a limited extent—almost exclusively for the W grades. Normalizing, for most steels, is used to refine the coarse grain size that results from hot working. However, most of the highly alloyed tool steels are slow cooled after hot working and then annealed. Normalizing of the highly alloyed grades after forging would likely cause cracking and would accomplish nothing.

The P grades shown in Table 1 (generally very low in carbon) are most often hardened by case hardening—carburizing or nitriding (see Chapter 10).

Table 1 Classification and approximate compositions of some principal types of tool steels
Source: AISI Steel Products Manual — Tool Steels, 1970.

AISI	UNS	C	Mn	Si	Cr	V	W	Mo	Co	Ni
Water-hardening tool steels										
W1	T72301	0.60-1.40 (a)	...	...	...	...	...	...	...	...
Shock-resisting tool steels										
S1	T41901	0.50	...	...	1.50	...	2.50	...	...	...
S5	T41905	0.55	0.80	2.00	...	...	...	0.40	...	...
Oil-hardening cold work tool steels										
O1	T31501	0.90	1.00	...	0.50	...	0.50	...	...	...
O2	T31502	0.90	1.60	...	...	...	...	...	...	...
Air-hardening medium-alloy cold work tool steels										
A2	T30102	1.00	...	...	5.00	...	...	1.00	...	...
A3	T30103	1.25	...	...	5.00	1.00	...	1.00	...	...
High-carbon high-chromium cold work steels										
D2	T30402	1.50	...	...	12.00	1.00	...	1.00	...	...
D5	T30405	1.50	...	...	12.00	...	...	1.00	3.00	...
Mold steels										
P2	T51602	0.07	...	...	2.00	...	...	0.20	...	...
P4	T51604	0.07	...	...	5.00	...	...	0.75	...	0.50
Chromium hot work tool steels										
H11	T20811	0.35	...	...	5.00	0.40	...	1.50	...	...
H12	T20812	0.35	...	...	5.00	0.40	1.50	1.50	...	...
H13	T20813	0.35	...	...	5.00	1.00	...	1.50	...	...
Tungsten hot work tool steels										
H21	T20821	0.35	...	...	3.50	...	9.00	...	...	...
H26	T20826	0.50	...	...	4.00	1.00	18.00	...	...	...
High speed tool steels										
T1	T12001	0.75 (a)	...	...	4.00	1.00	18.00	...	...	...
T8	T12008	0.75	...	...	4.00	2.00	14.00	...	5.00	...
T15	T12015	1.50	...	...	4.00	5.00	12.00	...	5.00	...
M1	T11301	0.80 (a)	...	...	4.00	1.00	1.50	8.00	...	...
M2	T11302	0.85-1.00 (a)	...	...	4.00	2.00	6.00	5.00	...	...
M10	T11310	0.85-1.00 (a)	...	...	4.00	2.00	...	8.00	...	...
M36	T11336	0.80	...	...	4.00	2.00	6.00	5.00	8.00	...
M44	T11344	1.15	...	...	4.25	2.00	5.25	6.25	12.00	...

(a) Available with different carbon contents as specified

Apparent Hardness Versus Actual Hardness

At this time, before discussing the heat treating tool steels, the reader should clearly understand the difference between "apparent" and "actual" hardness.

The general principles of hardness and hardness measurement were discussed in Chapter 3; it was indicated that microhardness testing is often required to reveal true conditions.

For most carbon and alloy steels, the apparent, or measured, hardness as indicated by test instruments is generally accurate and represents nearly true conditions because the structures of carbon and alloy steels are generally homogeneous.

For highly alloyed steels, such as high-carbon stainless and many tool steels, the structures are not necessarily homogeneous. For example, the structure of hardened W1 tool steel (Fig. 1) shows a high degree of homogeneity and measures 64 HRC, thus registering nearly true conditions.

Now let us examine the hardness of a highly alloyed tool steel such as D2. Figure 2 shows the structure of D2 in the quenched and tempered condition. The dark matrix is tempered martensite, not unlike the matrix of W1 shown in Fig. 1. Likewise, the registered hardness of the two is essentially the same— 62 HRC or about 760 on the Knoop scale. However, an entirely different condition exists in Fig. 2—a substantial addition of undissolved complex alloy carbides (white constituent). Measurements of these individual carbides determined by microhardness testing show readings of 2000 HK or higher. In testing with instruments such as the Rockwell tester, the carbide particles are simply "pushed" away and have little effect on the reading. Therefore, apparent or measured hardness is the hardness value provided by testing with conventional hardness testers, whereby the actual or true hardness takes into account the hardness of the individual constituents.

The practical difference between the two microstructures in Fig. 1 and 2 is their varying resistance to abrasion or wear. For example, sheet metal forming dies made of D2 have been known to outlast W1 counterparts by a factor of 5 to 1, simply because of the difference in resistance to wear of the two structures.

Heat Treating Practice for Specific Groups

The American Iron and Steel Institute lists a total of three W steel compositions, but W1 (Table 1) is by far the most widely used. Recommended practice for annealing and hardening W1 is given in Table 2. Except for extremely thin sections (generally under 3/16 in. or 4.76 mm), the W steels must be quenched in water or brine to attain full hardness. All W tool steels can be hardened to about 65 HRC at the surface, but their hardenability is very low, so that this high hardened zone is generally very shallow. However, a hard exterior and a softer core is desirable for many applications like punches.

Even though all of the W steels have relatively low hardenability, these grades usually are available as shallow, medium, or deep hardening; this property is controlled by the manufacturer.

Shock-Resisting Tool Steels. Five steels are listed by AISI under the symbol S, although only the two most widely used grades are shown in Table 1 (S1 and S5).

There is a considerable difference between the compositions of S1 and S5, as shown in Table 1. However, they are both intended for similar applications—applications that require extreme toughness such as punches, shear knives, and air-operated chisels.

These steels have sufficient hardenability so that full hardness can be attained by oil quenching. They must be tempered to at least 350 or 400 °F (175 to 205 °C) and preferably higher if some hardness can be sacrificed (see general section on tempering later in this chapter).

Details of recommended practice for annealing and hardening S1 and S5 are given in Table 2.

Oil Hardening Cold Work Tool Steels. Four O grades are listed by AISI, but only the two most widely used grades (O1 and O2) are listed in Table 1. Of these, O1 is used more frequently.

Heat treating practices for both grades are given in Table 3. As indicated, the high hardenabiliby of O1 is derived from the relatively high manganese content (1.0%) and from small additions of chromium and tungsten. Grade O2 depends on manganese content for hardenability. Both of these steels have far greater hardenability than W1 and thus are used where deep hardening by oil quenching is desired.

Air-Hardening, Medium-Alloy Cold Work Tool Steels. The cold work tool steels listed under the letter A cover a wide range of carbon and alloy contents, but all have high hardenability and exhibit a high degree of dimensional stability in heat treatment. A total of 10 different compositions are listed by AISI, although A2 and A3 only are listed in Table 1. These are, by far, the two most widely used for the A group. As shown in Table 1, A3 has a higher carbon content than A2 and also contains 1.0% V, thus some vanadium carbides are formed which provides A3 with superior wear resistance. Each of these grades contains 5.0% Cr so that some free chromium carbide characterizes the microstructures of these grades as shown for D2 in Fig. 2. Also, both A2 and A3 are sufficiently high in alloy content to develop some secondary hardening (see the general section on tempering later in this chapter).

Specific heat treating procedures are given for A2 and A3 in Table 3. As indicated, recommended heat treating procedures are almost the same for these two grades.

Table 2 Heat treating practice for the W and S grade tool steels
Source: AISI Products Manual—Tool Steels, 1970.

Treatment	Type W1 (T72301)	Type S1 (T41901)	Type S5 (T41905)
Annealing			
Temperature, °F........................	1360-1450	1450-1500	1425-1475
Rate of cooling, °F max per hour.........	40	40	25
Typical annealed hardness, HB	156-201	183-229	192-229
Hardening			
Rate of heating........................	Slowly	Slowly	Slowly
Preheat temperature, °F	(a)	1200	1400
Hardening temperature, °F..............	1400-1550	1650-1750	1600-1700
Time at temperature, min	10-30	15-45	5-20
Quenching medium	Brine or water	Oil	Oil
Tempering			
Tempering temperature, °F	350-650	400-1200	350-800
Approximate tempered hardness, HRC	64-50	58-40	60-50

(a) For large tools and tools having intricate sections, preheating at 1050-1200 °F is recommended.

Table 3 Heat treating practice for the O and A grade tool steels
Source: AISI Steel Products Manual—Tool Steels, 1970.

Treatment	Type O1 (T31501)	Type O2 (T31502)	Type A2 (T30102)	Type A3 (T30103)
Annealing				
Temperature, °F........................	1400-1450	1375-1425	1550-1600	1550-1600
Rate of cooling, °F maximum per hour....	40	40	40	40
Typical annealed hardness, HB	183-212	183-217	201-235	207-229
Hardening				
Rate of heating........................	Slowly	Slowly	Slowly	Slowly
Preheat temperature, °F	1200	1200	1450	1450
Hardening temperature, °F..............	1450-1500	1400-1475	1700-1800	1750-1850
Time at temperature, min	10-30	5-20	20-45	25-60
Quenching medium	Oil	Oil	Air	Air
Tempering				
Tempering temperature, °F	350-500	350-500	350-1000	350-1000
Approximate tempered hardness, HRC	62-57	62-57	62-57	65-57

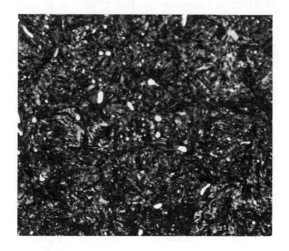

Fig. 1 Microstructure of W1 tool steel after brine quenching from 1450 °F (790 °C) and tempering at 350 °F (175 °C). Dark matrix is tempered martensite. A few undissolved particles of carbide are visible (white constituent). Hardness is 64 HRC. Source: *Metals Handbook*, Vol 7, 8th edition, American Society for Metals, p 102, 1972.

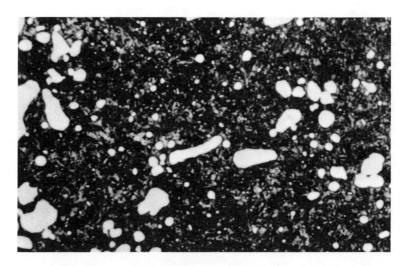

Fig. 2 Microstructure of D2 tool steel after air cooling from 1800 °F (980 °C) and tempering at 1000 °F (540 °C). Hardness is approximately 62 HRC. Dark matrix is tempered martensite with a dispersion of very hard carbide particles (white). See text for discussion. Source: *Metals Handbook*, Vol 7, 8th edition, American Society for Metals, p 112, 1972.

High-Carbon, High-Chromium Cold Work Tool Steels. The American Iron and Steel Institute lists five different compositions in the D group. They are all high in carbon (1.5% or higher) and contain a nominal amount (12.0%) of chromium. Compositions for the two most popular grades are given in Table 1. Note that D2, which is by far the most widely used D-type steel, also contains 1.0% V and 1.0% Mo; thus, it becomes fully hardened by air cooling from the austenitizing temperature, forming a martensitic matrix with large amounts of excess carbide as shown in Fig. 2.

As a rule, D2 tool steel is frequently used for dies employed for cold working operations, although it does have some resistance to softening from elevated temperatures.

Grade D5 is similar to D2 in composition, except that it contains 3.0% Co instead of vanadium. This compositional difference lowers its abrasion resistance, but the cobalt addition increases its resistance to softening from heat. Consequently, D5 often is used for hot forming tools.

Heat treating procedures for both D2 and D5 are presented in Table 4.

Mold Steels (P Grades). Of the seven mold steel compositions listed by AISI, compositions of two of these grades are listed in Table 1. Heat treatments commonly used for each are given in Table 5. Both of these steels are very low in carbon content, which is intentional to permit forming of plastic molds by hubbing. As shown in Table 5, case hardening (carburizing, see Chapter 10) is the most commonly used heat treatment. Grade P2 is widely used for molding a variety of plastics. With the addition of 5.0% Cr, P4 has greater resistance to deterioration from heat, so that it usually is used for molding plastics that require higher temperatures. Grade P4 also is used for dies for die casting of zinc.

Hot Work Steels (H Grades). The total AISI list is comprised of 13 different grades; compositions for the most widely used grades are given in Table 1. Heat treating procedures for each are given in Table 6. Note that they are all characterized by medium carbon content and that they are all highly alloyed, although their total alloy contents are by no means equal.

The chromium types (H11, H12, and H13) are used principally for hot dies and hot die inserts, blades for hot shearing, and dies for die casting of aluminum. The tungsten grades are capable of withstanding higher temperatures than the chromium types. H121 and H26 are used extensively for hot extrusion tools and dies for die casting of copper alloys.

High Speed Steels (T and M Grades). The total AISI listing of high speed steels comprises a family of 18 grades. Compositions of eight of the most important types are given in Table 1.

High speed steels are so named because of their resistance to heat and abrasion. Also, one of their principal uses is for cutting other metals at high speeds.

Table 4 Heat treating practice for grade D2 and D5 tool steels

Source: AISI Steel Products Manual—Tool Steels, 1970.

Treatment	Type D2 (T30402)	Type D5 (T30405)
Annealing		
Temperature, °F	1600-1650	1600-1650
Rate of cooling, °F maximum per hour	40	40
Typical annealed hardness, HB	217-255	223-255
Hardening		
Rate of heating	Very slowly	Very slowly
Preheat temperature, °F	1500	1500
Hardening temperature, °F	1800-1875	1800-1875
Time at temperature, min	15-45	15-45
Quenching medium	Air	Air
Tempering		
Tempering temperature, °F	400-1000	400-1000
Approximate tempered hardness, HRC	61-54	61-54

Table 5 Heat treating practice for grade P2 and P4 mold steels

Source: AISI Steel Products Manual—Tool Steels, 1970.

Treatment	Type P2 (T51602)	Type P4 (T51604)
Annealing		
Temperature, °F	1350-1500	1600-1650
Rate of cooling, °F maximum per hour	40	25
Typical annealed hardness, HB	103-123	116-128
Hardening		
Carburizing temperature, °F	1650-1700	1775-1825
Hardening temperature, °F	1525-1550	1775-1825
Time at temperature, min	15	15
Quenching medium	Oil	Air
Tempering		
Tempering temperature, °F	350-500	350-900
Approximate tempered hardness, HRC	64-58	64-58

Table 6 Heat treating practice for hot work tool steels
Source: AISI Steel Products Manual—Tool Steels, 1970.

Treatment	Type H11 (T20811)	Type H12 (T20812)	Type H13 (T20813)	Type H21 (T20821)	Type H26 (T20826)
Annealing					
Temperature, °F	1550-1650	1550-1650	1550-1650	1600-1650	1600-1650
Rate of cooling, °F maximum per hour	40	40	40	40	40
Typical annealed hardness, HB	192-235	192-235	192-229	207-235	217-241
Hardening					
Rate of heating	Moderately from preheat	Moderately from preheat	Moderately from preheat	Rapidly from preheat	Rapidly from preheat
Preheat temperature, °F	1500	1500	1500	1500	1600
Hardening temperature, °F	1825-1875	1825-1875	1825-1900	2000-2200	2150-2300
Time at temperature, min	15-40	15-40	15-40	2-5	2-5
Quenching medium	Air	Air	Air	Air, oil	Salt, air, oil
Tempering					
Tempering temperature, °F	1000-1200	1000-1200	1000-1200	1100-1250	1050-1250
Approximate tempered hardness, HRC	54-38	55-38	53-38	54-36	58-43

High speed steels commonly are classified in two ways—by composition and by the amount of cobalt they contain. Classification by composition involves the amount of tungsten(T grades), or molybdenum(M grades) as the principal alloying element. The second method of classification is determined by whether or not they contain substantial amounts of cobalt. Those grades such as T1 and M2 are commonly known as general-purpose types, while those that contain relatively large amounts of cobalt are called extra-heavy duty or super-high speed steels. They are used for applications that are extremely demanding in terms of operating at elevated temperature. The use of these "super" grades is relatively small compared to the total usage of high speed steels. In fact, it was recently estimated that 80% of the total amount of high speed steel produced was comprised of M1, M2, and M10.

Heat treating practice for the eight compositions of high speed steels shown in Table 1 are given in Tables 7 and 8. High austenitizing temperatures are required to dissolve sufficient amounts of the very difficult-to-dissolve complex carbides of tungsten, molybdenum, vanadium, and chromium. The quenching medium may be either oil or air. Even though all of the high speed steels have hardenability that is sufficient to attain full hardness by air

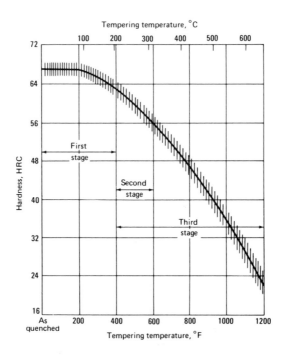

Fig. 3 Hardness versus tempering temperature for W1 tool steel after brine quenching from 1450 °F (790 °C). Source: *Heat Treater's Guide: Standard Practices and Procedures for Steel*, American Society for Metals, p 262, 1981.

cooling, oil quenching may be used to minimize scaling. One approach that is often used for heat treating tools made from high speed steels is to oil quench "to black" from the austenitizing temperature, then to cool to near room temperature in still air. If the tools have been austenitized in a molten salt bath, they should be given a "quick quench," which is actually a rinse at approximately 1100 °F (595 °C) in a water-soluble salt and then air cooled. If a high-temperature salt is allowed to solidify on the tool, it is almost impossible to remove.

High speed steels usually are tempered at relatively high temperatures, and double or multiple tempering is essential.

Tempering of Tool Steels

All austenitized and quenched tool steels should be tempered to at least 300 °F (150 °C) for a length of time that ensures that tools have been heated throughout the section thickness involved (usually a minimum time of 1 h per inch of maximum section). From here on, tempering time and temperature (mainly temperature) depend greatly on the grade of steel and/or its end use.

Three important rules to establish tempering techniques for tool steels are: (1) use a temperature that is as high as can be tolerated in terms of loss of hardness; (2) always temper the steel at a temperature that is at least as high,

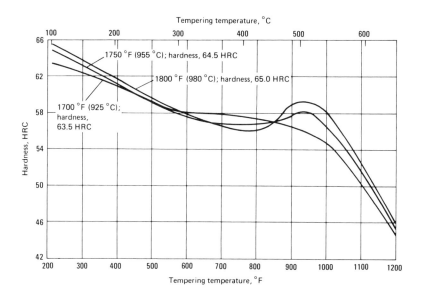

Fig. 4 Hardness versus tempering temperature for A2 tool steel, austenitized as shown and air cooled. Source: *Heat Treater's Guide: Standard Practices and Procedures for Steel*, American Society for Metals, p 288, 1981.

and preferably slightly higher, than the maximum temperature to which it will be subjected in service; and (3) never allow the quenched part to quite reach room temperature before beginning the tempering operation. Careful adherence to the third rule becomes more important as carbon and/or alloy content increases. While the workpiece must be allowed to complete its transformation, when the workpiece is still "just a little hot" to the touch (about 125 °F, or 50 °C) is generally the ideal time to place it in the tempering furnace.

Tempering Temperature Versus Hardness. For the W and O grades of tool steels, hardness after tempering can be expected to decrease gradually as tempering temperature increases (Fig. 3). To a great extent, this also holds true for the S grades.

For the more highly alloyed A grades, the conditions change, and secondary hardening (humps in the curves) becomes evident. The mechanism of secondary hardening was explained in Chapter 7 in connection with heat treating of high-carbon stainless steels. Specific curves for A2 after austenitizing are given in Fig. 4.

For the D grades, the tempering temperature versus hardness curves are similar to Fig. 4, except that secondary hardening is more pronounced for the more highly alloyed grades. Often, the hardness at the peak of secondary hardening equals the as-quenched hardness.

Tempering curves for chromium-type hot work grades (H11, H12, and H13) generally are similar to those shown in Fig. 4. Because of their lower

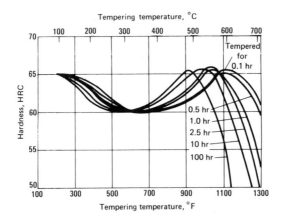

Fig. 5 Hardness versus tempering temperature for M2 high speed tool steel after austenitizing at 2225 °F (1220 °C) and tempering for time periods and temperatures as shown. Source: *Heat Treater's Guide: Standard Practices and Procedures for Steel,* American Society for Metals, p 377, 1981.

Table 7 Heat treating practice for tungsten high speed steels
Source: AISI Steel Products Manual—Tool Steels, 1970.

Treatment	Type T1 (T12001)	Type T8 (T12008)	Type T15 (T12015)
Annealing			
Temperature, °F.....................	1600-1650	1600-1650	1600-1650
Rate of cooling, °F maximum per hour..	40	40	40
Typical annealed hardness, HB	217-255	229-255	241-277
Hardening			
Rate of heating......................	Rapidly from preheat	Rapidly from preheat	Rapidly from preheat
Preheat temperature, °F	1500-1600	1500-1600	1500-1600
Hardening temperature, °F.............	2300-2375	2300-2375	2200-2300
Time at temperature, min	2-5	2-5	2-5
Quenching medium	Oil, air, salt	Oil, air, salt	Oil, air, salt
Tempering			
Tempering temperature, °F	1000-1100	1000-1100	1000-1200
Approximate tempered hardness, HRC ..	65-60	65-60	68-63

carbon content, however, initial hardness (as-quenched) is lower. The tungsten-type hot work steels (H21 and H26) behave more like the high speed steels during tempering.

Hardness versus tempering temperature curves for the most widely used grade of high speed tool steel (M2) are presented in Fig. 5. As shown, hardness drops at first; that is, after tempering at about 550 °F (290 °C), hardness is lower than the initial hardness by approximately five points HRC. Secondary hardening then begins, and tempering between 900 to 1100 °F (480 to 595 °C) produces hardnesses that are approximately the same as the initial hardness. Similar tempering temperature versus hardness patterns prevail for all of the high speed steels. Consequently, a tempering temperature of 1050 °F (565 °C) is almost universally used for hardware items made from high speed steels.

Double and Multiple Tempering. While double tempering—tempering for a given time at an established temperature, cooling to near room temperature, and then tempering again at the established temperature—is never wrong, it is usually considered a waste of time and energy for steels that exhibit no sign of secondary hardening (Fig. 3).

Steels that do exhibit secondary hardening (Fig. 4 and 5) should at least be double tempered. Triple tempering (three complete cycles) is frequently used for high cobalt grades like M44. In fact, some heat treaters employ as many as five cycles for the highly alloyed grades in an attempt to "reach for the last bit" of transformation for the austenite.

Table 8 Heat treating practice for molybdenum high speed steels
Source: AISI Steel Products Manual—Tool Steels, 1970.

Treatment	Type M1 (T11301)	Type M2 (T11302)	Type M10 (T11310)	Type M36 (T11336)	Type M44 (T11344)
Annealing					
Temperature, °F	1500-1600	1600-1650	1500-1600	1600-1650	1600-1650
Rate of cooling, °F maximum per hour	40	40	40	40	40
Typical annealed hardness, HB	207-235	212-241	207-255	235-269	248-285
Hardening					
Rate of heating	Rapidly from preheat	Rapidly from preheat	Rapidly from preheat	Rapidly from preheat	Rapidly from preheat
Preheat temperature, °F	1350-1550	1350-1550	1350-1550	1350-1550	1350-1550
Hardening temperature, °F	2150-2225	2175-2250	2150-2225	2225-2275	2190-2240
Time at temperature, min	2-5	2-5	2-5	2-5	2-5
Quenching medium	Oil, air, salt	Oil, air, salt	Oil, air, salt	Oil, air, salt	Oil, air, salt
Tempering					
Tempering temperature, °F	1000-1100	1000-1100	1000-1100	1000-1100	1000-1100
Approximate tempered hardness, HRC	65-60	65-60	65-50	65-60	70-62

Chapter 9

Heat Treating
of Cast Irons

The term "cast iron" as covered in this treatise includes gray, white, ductile, and malleable irons. However, heat treatment of malleable iron is done largely by the manufacturer. The use of malleable irons is gradually decreasing in favor of ductile irons. Therefore, heat treating procedures described herein will be confined to the higher tonnage irons—gray and ductile. For more complete information on all grades of cast irons including the compacted graphite and the special-purpose grades, see Ref 10, 17, and 18.

Essentially, cast iron is an alloy of iron, carbon, and silicon with a total carbon content much higher than that found in steel (see the right side of the iron-carbon equilibrium diagram shown in Fig. 1). Silicon is an important control element in cast iron and, therefore, must be given full consideration. The iron-graphite phase diagram shown in Fig. 1 represents a modification of the iron-cementite diagram presented in Chapter 2. One principal difference between the two diagrams is that the composition shown in Fig. 1 of this chapter contains 2.5% Si (a typical amount in gray and ductile irons). To interpret Fig. 1, use the same technique as described for use of the iron-cementite diagram shown in Chapter 2—the intersection of any line drawn from the vertical and horizontal axes shows the phase or phases present in that area, as a function of temperature and carbon content.

Further, in cast irons, carbon is present in excess of the amount that is retained in solid solution in austenite at the eutectic temperature. This excess carbon is in the form of graphite, so that cast irons are not only alloys; they are also considered composites or mixtures. As a rule, the range of total carbon in cast irons is approximately 1.75 to 4.0%, with varying amounts of silicon up to about 2.8%. Some special-purpose grades of cast iron contain up to about 12.0% Si. In addition, commercial grades commonly contain manganese in the range of approximately 0.40 to 0.90% and sometimes higher. Sulfur content is generally less than 0.15%, and phosphorus is usually within the range of 0.02 to 0.90%. Alloying elements such as nickel, chromium, molybdenum, vanadium, and copper—singly or in combination—may be added to develop specific properties, much the same as for steels.

The ductility of gray irons in the as-cast condition is very low, ranging from virtually zero up to 2% in terms of elongation. Ductile cast irons, however, possess ductile properties that approach those of steels with similar microstructures. Heat treated malleable irons also possess considerable ductility.

Differences Among Types of Cast Irons

The principal basis for distinguishing among various types of cast irons is to determine the form in which carbon exists. The two vital factors in determining the carbon behavior are chemical composition and cooling rate. The composition determines whether carbon will be initially stable as a metal carbide (white cast iron) or would prefer to exist as graphite in flake form (gray cast iron), in spheroidal form (ductile cast iron), as a mixture of spheroidal and vermicular graphite shapes (compacted graphite iron), or as metal carbide temporarily and later forming nodules of graphite (malleable cast iron) when annealed at 1500 to 1700 °F (815 to 925 °C). Table 1 lists typical compositions for several classes of cast irons. For further information on irons other than conventional gray irons and ductile irons, see Ref 10 and 17.

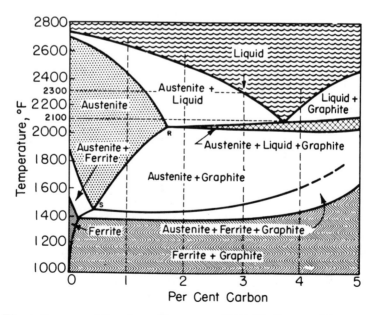

Fig. 1 Iron-graphite phase diagram at 2½% Si. Source: Metals Engineering Institute Course 17, Lesson 8, American Society for Metals, p 2, 1981.

Table 1 Typical compositions of major classes of cast irons

Source: Metals Engineering Institute, Course 17, Lesson 8, American Society for Metals, p 2, 1981.

Irons	Total C (b)	Composition (a), %				
		Si	Mn	P	S	Other
Gray cast iron						
Ordinary grade	**3.4**	**2.0**	0.6	0.20	0.10	...
High-strength grade	**3.0**	**1.5**	0.8	0.20	0.10	...
White cast iron	**2.0-3.5** (all combined)	**0.5-1.5**	0.5	0.2-0.4	0.10	...
Malleable cast iron...............	**2.3**	**1.0**	0.4	0.20	0.10	...
Ductile cast iron						
Ferritic	3.5	2.5	**0.2**	0.05	0.01	**0.04 Mg**
Pearlitic	3.5	2.5	**0.4**	0.05	0.01	**0.04 Mg**

(a) Important differences appear in boldfaced type. (b) Total carbon content (includes graphite plus combined carbon).

For cast irons with chemical compositions that place them between white and gray cast irons, cooling rate is the dominant variable. Rapid cooling favors the formation of white cast irons (chilled cast iron), whereas slow cooling favors gray cast irons. Intermediate cooling rates can result in the simultaneous formation of metal carbide and flake graphite (mottled cast irons). Mottled cast irons are simply a mixture of white and gray cast iron. The number of different irons that can exist is consequently almost infinite.

For the two general types of iron that are emphasized in this chapter—gray and ductile—the form in which the graphite particles exists has a distinct affect on mechanical properties in their as-cast as well as their heat treated condition. The composition of a ductile iron may be almost identical to that of a plain gray iron, and yet their mechanical properties are widely different because of the difference in graphite shape (see Fig. 2).

Carbon Equivalent

Before discussing heat treating techniques for cast irons, the reader should understand the concept of carbon equivalent.

Silicon and phosphorus considerably decrease the amount of carbon necessary to form eutectic composition with iron. This is important, because as eutectic composition is approached, the melting point decreases, fluidity increases, and strength of the iron decreases. This effect is measured by carbon equivalent (CE) and is of great importance in the heat treating response and mechanical properties of cast iron. The most commonly used formula for calculating CE is:

$$CE = \%C + \%Si/3 + \%P/3$$

Gray iron foundaries often keep phosphorus very low, so they delete it from the CE equation and simply state:

$$CE = \%C + \%Si/3$$

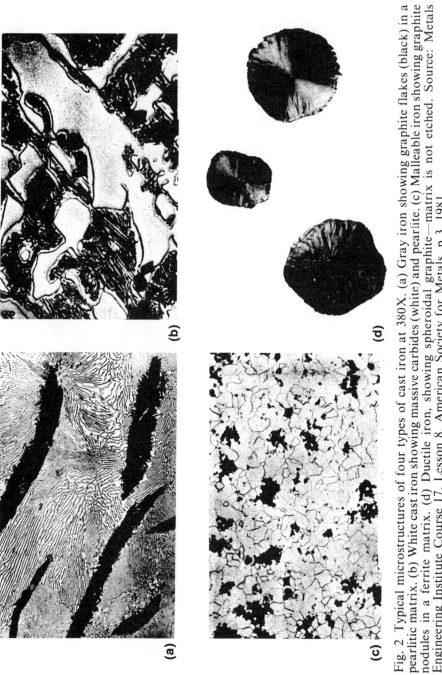

Fig. 2 Typical microstructures of four types of cast iron at 380X. (a) Gray iron showing graphite flakes (black) in a pearlitic matrix. (b) White cast iron showing massive carbides (white) and pearlite. (c) Malleable iron showing graphite nodules in a ferrite matrix. (d) Ductile iron, showing spheroidal graphite—matrix is not etched. Source: Metals Engineering Institute Course 17, Lesson 8, American Society for Metals, p 3, 1981.

As indicated in the sections that follow, cast irons having relatively low carbon equivalents respond better and more consistently to quench hardening because the combined carbon is higher.

Measuring Hardness of Cast Irons

Conventional hardness measurements on cast irons always indicate lower values than the true hardness of the matrix, because the graphite particles have essentially no hardness; thus, an indenter such as a Brinell ball or a Rockwell Brale cannot show true conditions because the area occupied by the graphite offers no resistance to an indenter. The amount of error depends not only on the total amount of carbon in the iron, but depends more on what percentage of the total carbon is in the form of graphite. Thus, if the precise composition of an iron is known (including the form in which the graphite exists) tables and charts are available that can provide a reasonably accurate conversion.

The situation is generally the same as discussed in Chapter 8, except that extremely hard microconstituents were involved; for cast irons, soft constituents cause inaccuracies in hardness testing. In either case, microhardness testing is the only approach for determining true hardness.

As an example, comparative hardness readings obtained on ten quenched and tempered ductile irons are given in Table 2. Observed HRC readings range from 3.8 to 8.3 points lower than those converted from microhardness readings.

Figure 3 shows the relation between observed HRC readings and those converted from microhardcness values for five gray irons of different carbon equivalents. Hardness measurements were taken at two laboratories after quenching and after tempering of each iron. The data in Fig. 3 show why the

Table 2 Comparative hardness values for quenched and tempered ductile irons

Source: Metals Handbook, Vol 4, 9th edition, American Society for Metals, p 526, 1981.

Iron	HB(a)	HRC converted from HB(b)	Observed HRC(c)	Microhard-ness, HV(d)	HRC converted from HV(b)	HRC converted from HV minus observed HRC
1	415	44.5	44.4	527	50.9	6.5
2	444	47.2	45.0	521	50.6	5.6
3	444	47.2	45.7	530	51.1	5.4
4	444	47.2	47.6	593	54.9	7.3
5	461	48.8	46.7	595	55.0	8.3
6	461	48.8	48.3	560	53.0	4.7
7	461	48.8	49.1	581	54.2	5.1
8	477	50.3	49.6	572	53.7	4.1
9	477	50.3	50.1	618	56.2	6.1
10	555	55.6	53.4	637	57.2	3.8

(a) Average of three readings for each iron. (b) Values based on SAE-ASM-ASTM hardness conversions for steel. (c) Average of five readings for each iron. (d) Average of a minimum of five readings for each iron; 100-kg load.

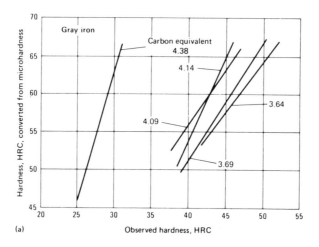

Fig. 3 Relations between observed and converted hardness values for gray iron. Source: *Metals Handbook*, Vol 4, 9th ed., American Society for Metals, p 527, 1981.

observed values obtained by conventional hardness testing may be misleading and help to explain the good wear resistance of gray irons with apparently low hardness. Note that there is a correlation with carbon equivalent for all five irons tested and that the discepancy between observed and converted hardness values diminishes at the lower level of carbon equivalent.

Heat Treatments for Gray Irons

The types of heat treatments used for gray irons include annealing, normalizing, heating for hardening by quenching, tempering, austempering, martempering, flame and induction hardening, and stress relieving (see Chapters 5, 6, and 10 for discussions on these special treatments).

In general, the principles involved for heat treating of gray irons do not differ greatly from those used for carbon and alloy steels. One significant difference is the care that must be taken to avoid cracking of gray iron castings. They are inherently brittle and are susceptible to cracking from thermal shock.

The cardinal rule for heating gray iron, regardless of whether it is to be normalized, annealed, or austenitized for quenching, is to heat slowly. Any gray iron castings should be heated slowly through the lower temperature range. Above a range of 1100 to 1200 °F (595 to 650 °C), heating may be as rapid as desired. In fact, time may be saved by heating the casting slowly to about 1200 °F (650 °C) in one furnace and then transferring it to a second furnace and bringing it rapidly up to the austenitizing temperature.

Annealing of Gray Irons

With the possible exception of stress relieving, annealing is the heat treatment most frequently applied to gray irons. Annealing of gray irons consists of heating it to a temperature that is high enough to soften it, thereby improving its machinability. This is frequently the main reason for annealing.

Up to approximately 1100 ° F (595 ° C), the effect of temperature on the structure and hardness of gray irons is insignificant. As the temperature increases above 1100 ° F (595 ° C), the rate at which iron carbide decomposes to ferrite plus graphite increases markedly reaching a maximum at the lower transformation temperature (about 1400 ° F or 760 ° C for unalloyed or low-alloy iron).

Gray irons usually are subjected to one of three annealing treatments, each of which involves heating to a different range of temperature. These treatments are the ferritizing anneal, the medium (or full) anneal, and the graphitizing anneal.

Ferritizing Annealing. This process may be used for unalloyed or low-alloy gray iron of normal composition, when the only result desired is the conversion of pearlitic carbide to ferrite and graphite for improved machinability. It generally is unnecessary to heat the casting above 1400 ° F (760 ° C).

For most gray irons, a ferritizing annealing temperature between 1300 and 1400 ° F (705 and 760 ° C) is recommended. Precise temperatures within this range depend on the exact composition of the iron. When machining properties are of primary importance, it is advisable to anneal a number of samples at various temperatues between 1300 and 1400 ° F (705 and 760 ° C) to determine the temperature that yields the desired final hardness. At temperatures between 1300 and 1400 ° F (705 and 760 ° C), holding time varies with chemical composition and may be as short as 10 min for thin sections of unalloyed cast irons. If an unusually slow rate of cooling is used, the time at temperature may be reduced further.

In ferritizing annealing, the cooling rate is seldom of great importance. If the stress relief that automatically occurs during annealing is to be retained, however, a maximum cooling rate of 200 ° F/h (110 ° C/h) is recommended.

While a ferritizing anneal greatly increases machinability of a gray iron, a considerable loss in hardness and stength results.

Full annealing is also called medium annealing, because the results lie somewhere between the ferritizing and graphitizing anneals. Full annealing usually is performed by heating in the range of 1450 to 1650 ° F (790 to 900 ° C).

This treatment is used when a ferritizing anneal would be ineffective because of the high alloy or high combined carbon content of a particular iron. However, it is recommended that the effect of temperature at or below 1400 ° F (760 ° C) be tested before a higher annealing temperature is adopted as part of a standard procedure.

In full annealing, holding times comparable to those of the ferritizing anneal are usually employed. However, when the high temperatures of full annealing are used, the casting must be cooled slowly through the transformation range from about 1450 to 1250 °F (790 to 675 °C).

In general, cast irons that have been subjected to full annealing are still lower in hardness and/or strength compared with irons that were annealed by ferritizing. Full annealing is not practiced extensively because of the loss of strength, which can be regained only by further heat treatments that are time-consuming and expensive.

As a rule, the correlation of casting design and composition should be such that full annealing should not be required. This refers specifically to section thickness. With some grades of iron—namely those with a low carbon equivalent and/or those containing chromium or molybdenum—the minimum section thickness becomes extremely critical. When section thicknesses are thin, there is a danger of white iron forming (see Fig. 2b). This constituent is nearly impossible to machine; subsequent annealing is required to decompose the white iron into its constituents—ferrite and graphite. It can thus be concluded that, if the castings are to be machined, white iron should not be allowed to form; breaking down white iron with heat simply degrades the properties of the casting.

Graphitizing Anneal. The graphitizing anneal represents the ultimate in annealing of gray iron—not only in decomposition of chilled iron or massive carbides, but it also generally reduces strength to a minimum. If the microstructure of a gray iron contains massive carbides or chilled areas, higher annealing temperatures may be needed to meet acceptable machinability requirements. The graphitizing anneal may simply convert massive carbide to pearlite and graphite. In some applications, it may be desired to carry decomposition all the way to a ferrite-graphite structure for maximum machinability. Here again, producing irons that require drastic annealing is usually accidental and should be avoided. However, in practice, such conditions often occur, and the ultimate in annealing is then required.

To break down the massive carbide with reasonable speed, temperatures of at least 1600 °F (870 °C) are required. With each additional 100 °F (55 °C) increment in holding temperature, the rate of carbide decomposition doubles; consequently, it is general practice to employ holding temperatures of 1650 to 1750 °F (900 to 954 °C). However, at 1700 °F (925 °C) and above, the phosphide eutectic present in irons containing 0.10% P or more may melt.

The holding time at temperature may vary from a few minutes to several hours. The chill (white iron) in some high-silicon, high-carbon irons can be eliminated in as little as 15 min at 1720 °F (940 °C). In all applications, unless a controlled-atmosphere furnace is used, the time at temperature should be as short as possible. This is because at these high temperatures, gray irons are susceptible to scaling.

The cooling rate chosen depends on the final use of the iron. If the principal object of the treatment is to break down carbides and if the retention of maximum strength and wear resistance is desired, the casting should be air cooled from the annealing temperature to about 1000 °F (540 °C) to promote the formation of a pearlitic structure. If maximum machinability is the object, the casting should be furnace cooled to 1000 °F (540 °C), and special care should be taken to cool slowly through the transformation range. In both instances, cooling from 1000 °F (540 °C) to about 550 °F (290 °C) at not more than 200 °F/h (110 °C/h) is recommended to avoid formation of residual stresses.

Normalizing of Gray Irons

Gray irons are normalized by being heated to a temperature above the transformation range, held at this temperature for a period of about 1 h for each 1 in. (25 mm) of maximum section thickness, and cooled in still air to room temperature. Normalizing may be used to enhance mechanical properties such as hardness and tensile strength. Also, it can restore as-cast properties that have been modified by another heating process such as graphitizing or the preheating and postheating associated with repair welding.

The temperature range for normalizing gray iron is approximately 1625 to 1700 °F (885 to 925 °C). Heating temperature has a marked effect on microstructure and on mechanical properties such as hardness and tensile strength. Because temperatures of 1500 °F (815 °C) and higher are above the transformation temperature, the matrix of the as-cast iron is converted to austenite on heating and is transformed during air cooling to ferrite-carbide

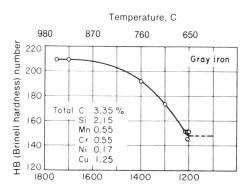

Fig. 4 Hardness after normalizing. Effect of temperature at start of air cooling on hardness of normalized gray iron rings 4¾ in. outside diameter, 1½ in. high and 1/12 in. thick. Source: Metals Engineering Institute Course 17, Lesson 8, American Society for Metals, p 6, 1981.

aggregates. These vary in fineness depending on the maximum temperature (the normalizing temperature) and the alloy content. For the alloy iron, the higher the normalizing temperature the harder and stronger the normalized structure. For the unalloyed iron, all normalizing temperatures produce the same hardness and strength, and all structures produced are generally softer than the as-cast material. Thus, normalizing is a hardening process for alloy irons and a softening process of unalloyed gray irons.

Some control of hardness can be exercised during normalizing by allowing castings to cool in the furnace to a temperature below the normalizing temperature. Figure 4 shows the results obtained with gray iron rings that were heated to 1750 °F (955 °C), then furnace cooled to different temperatures before being removed from the furnace and cooled in air. These data also indicate that annealing can be accomplished by cooling castings in the furnace to 1200 °F (650 °C) and then air cooling. However, if stress-free castings are desired, they should be cooled in the furnace to 550 °F (290 °C) before removal.

It can be concluded that normalizing of gray irons serves to restore as-cast properties. Further, if carbon equivalent is sufficiently low, normalizing causes as-cast properties to be exceeded.

Quenching and Tempering of Gray Irons

Gray irons can be hardened and tempered to improve their mechanical properties and their resistance to wear by using techniques similar to those used for steel.

Austenitizing Practice. In hardening gray irons, the casting generally is preheated to approximately 1000 °F (540 °C) and then is heated to a high

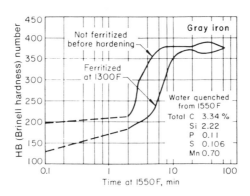

Fig. 5 Effect of austenitizing time on the hardness of quenched gray iron specimens 1¼ in. in diameter and ¾ in. thick. Source: Metals Engineering Institute Course 17, Lesson 8, American Society for Metals, p 7, 1981.

enough temperature to form austenite (see Fig. 5), held at that temperature for a sufficient length of time to affect solution of the desired amount of carbon, and then quenched at a rate suitable for that particular iron's composition.

The temperature to which the casting must be heated is determined by the transformation range of the particular gray iron of which it is made. A formula for determining the approximate transformation temperature of unalloyed gray iron is:

$$T_1 = 1345\ °F\ (730\ °C) + [50.4\ °F\ (28\ °C) \times \%Si] - [45\ °F\ (23\ °C) \times \%Mn]$$

where T_1 is the desired transformation temperature.

Chromium raises the transformation range of gray iron. In high-nickel, high-silicon irons, for example, each percent of chromium raises the transformation range by about 72 °F (39 °C). Nickel, on the other hand, lowers the transformation range.

Provided recommended limits are not exceeded, the higher the casting is heated above the transformation range the greater will be the amount of carbon dissolved in the austenite and the higher will be the hardness of the casting after quenching. In practice, temperatures as much as 175 °F (97 °C) higher than the calculated transformation temperature are used to ensure full austenitizing. However, excessively high temperatures should be avoided. Quenching from such high temperatures increases the danger of distortion and cracking and promotes retention of austenite.

Quenching. Cast irons are far more susceptible to cracking in quenching than most steels, which must be given full consideration in quenching. Oil is most frequently used as the medium for quenching gray iron castings. Oil at 180 to 220 °F (80 to 105 °C) can be used to minimize severity of quench when quenching castings of contrasting section size. Generally, water is not a satisfactory quenching medium for furnace-heated gray irons because it extracts heat so rapidly that distortion and cracking are likely in all but small parts of simple design. If water must be employed as a quenching liquid, a layer of oil is sometimes placed on top of the water to minimize thermal shock.

A casting of nonuniform section should be quenched in such a way that the heavier section enters the quenching bath first. During quenching, agitation is desirable because it ensures an even temperature distribution in the bath and improves quenching efficiency. Because as-quenched castings at room temperature are extremely sensitive to cracking, they should be removed from the quench bath as soon as their temperature falls to about 300 °F (150 °C) and tempered immediately.

Tempering. After quenching, castings are tempered at a preselected temperature which is well below the transformation temperature. The tempering temperature depends mainly upon the desired final hardness. Tempering time is generally 1 h per inch of section. As the quenched casting is tempered, the hardness decreases as tempering temperature is increased (see

Fig. 6). Also, as is indicated in Fig. 6, the rate of hardness decrease is influenced by the composition of the iron.

Furnaces used for tempering are most often the forced-air convection type, although salt baths also are used (see Chapter 4).

Heat Treating of Ductile Irons

In general, the same types of heat treatments used for gray irons can be successfully applied to ductile irons. As previously stated, composition of a given gray iron can have a ductile iron counterpart. Compositions can be nearly identical—the shape of the graphite particles is the principal difference.

It does follow that ductile irons are somewhat less susceptible to cracking during heat treating compared to gray iron counterparts.

Annealing. When maximum ductility and the best machinability are desired and high strength is not required, ductile iron castings generally are given a full anneal. The microstructure is converted to ferrite and spheroidal graphite. Manganese and phosphorus, and alloying elements such as chromium, nickel, copper, and molybdenum should be as low as possible if the best machinability is desired, because these elements retard the annealing process.

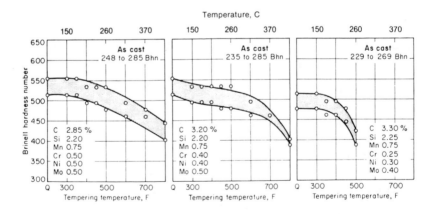

Fig. 6 Influence of alloy content on hardness of quenched and tempered gray iron test castings that were normalized to the same hardness range before being austenitized for hardening. Castings were quenched in oil from 1560 °F (850 °C). Source: Metals Engineering Institute Course 17, Lesson 8, American Society for Metals, p 9, 1981.

Working examples of two different annealing cycles are described as follows:

- Hold at 1650 to 1750 °F (900 to 955 °C), but furnace cool to 1200 °F (650 °C) so that cooling rate between 1450 and 1200 °F (790 and 650 °C) does not exceed 35 °F (19 °C) per hour. From this point, cooling rate is not critical, except that better stress relieving is obtained by slow cooling to below 800 °F (425 °C).
- A shorter, subcritical annealing cycle can be used when maximum impact properties are not required. In this procedure the castings are heated to 1300 °F (705 °C) and held for 5 h plus 1 h per inch of maximum section thickness. Then, castings are furnace cooled to at least 1100 °F (595 °C).

Normalizing can result in a considerable improvement in tensile properties, but with a large decrease in ductility. The microstructure obtained by normalizing depends on the composition of the casting and its cooling rate. The cooling rate depends on the mass of the casting. However, it also may be influenced by the temperature and movement of the surrounding air during cooling. Generally, normalizing produces a structure of fine pearlite, with graphite spheroids, if the metal is not too high in silicon and has at least a moderate manganese content. Heavier castings should contain alloying elements such as nickel, copper, molybdenum, and additional manganese for satisfactory normalizing. Lighter castings made of alloy iron may be martensitic after normalizing.

Normalizing should be followed by tempering to provide uniform hardness and resulting mechanical properties and to relieve residual stresses that develop when various section thicknesses of a casting are cooled in air at different rates. The effect of tempering on the hardness and tensile properties depends on the composition of the iron and on the hardness level that was obtained in normalizing. In general, pearlitic structures such as those resulting from normalizing soften less than the harder martensitic structures that are obtained by quenching.

Quenching and Tempering. A temperature of 1550 to 1700 °F (845 to 925 °C) normally is used for austenitizing commercial ductile iron castings and produces the highest as-quenched hardness. To minimize stresses, oil is preferred as a quenching medium, but water or brine may be used for simple shapes.

The influence of the austenitizing temperature on the hardness of water-quenched ½-in. (13-mm) cubes of ductile iron is shown in Fig. 7. These data show that the highest range of hardness (55 to 57 HRC) was obtained with austenitizing temperatures between 1550 and 1600 °F (845 and 870 °C). Specimens quenched from 1700 °F (925 °C) contained enough retained austenite to lower the hardness to 47 HRC (Fig. 7).

Time at austenitizing temperature also is important for obtaining full hardness. This was determined by first heating as-cast specimens in molten salt at 1600 °F (870 °C) and then quenching in water. Specimens that were heated for 2 min contained 30 to 35% ferrite in the microstructure and developed a hardness of 32 to 45 HRC. When similar specimens were held for 4 min, the retained ferrite decreased to 12 to 15% and hardness increased to 44 to 51 HRC. Further, when specimens were held for 10 min 1600 °F (870 °C), the ferrite had disappeared and hardness was fully developed (53 to 57 HRC).

Tempering. After quenching, ductile iron castings are usually tempered for 1 h plus 1 h per inch of maximum section thickness. The results in terms of hardness versus tempering temperature can be held to closer ranges than those in gray iron, although hardness values after tempering are generally similar to those shown for gray iron in Fig. 6 for similar compositions.

Stress Relieving of Gray and Ductile Cast Irons

Castings of gray or ductile iron may contain high-magnitude residual stresses, because cooling (and therefore contraction) proceeds at different rates throughout various sections of a casting. The resultant residual stresses cause distortion. In extreme cases, they even result in failure or cracking when they approach the ultimate strength of the material. The magnitudes of these stresses depend on the shape and dimensions of the casting, on the casting technique employed, on the composition and properties of the material cast, and on whether the casting has been stress relieved.

As a rule, a reasonably complete removal of residual stresses can be achieved by heating at 1000 or 1050 °F (540 to 565 °C). This temperature seldom decreases the hardness or strength of the as-cast condition.

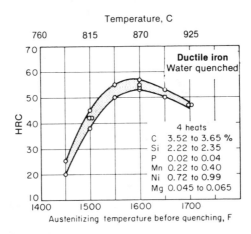

Fig. 7 Influence of austenitizing temperature on hardness of ductile iron. Each value represents the average of three hardness readings. Specimens (½-in. cubes) were heated in air for 1 h and water quenched. Source: Metals Engineering Institute Course 17, Lesson 8, American Society for Metals, p 19, 1981.

The rate of heating for stress relief depends on the shape of the part. Except for parts having thick and thin sections, rate of heating is not especially critical. When a batch-type furnace is employed, it is of the utmost importance that furnace temperature not exceed 200 °F (95 °C) at the time of loading. After the furnace is loaded, the heating rate may be fairly rapid. For example, it is common practice to heat to 1050 °F (565 °C) in about 3 h, hold at temperature for 1 h per inch of section, and cool to 600 °F (315 °C) in about 4 h before removing castings from the furnace and permitting them to cool in air. These conditions apply also to continuous furnaces in which the various zones can be controlled to avoid introducing additional thermal stress in the castings. For very precise machining tolerances, slower heating and cooling rates are generally specified.

Hardenability of Gray and Ductile Irons

The meaning of and method of determining hardenability for gray and ductile irons (carbon and alloy) are essentially the same as for steel (see Chapter 3). It must be emphasized that hardenability is that property of the iron which controls the depth of hardness which occurs when quenched from above the transformation temperature, and should not be confused with hardness. Maximum attainable hardness is governed principally by the content of combined carbon. However, cast irons contain two forms of carbon—combined and graphitic. At the austenitizing temperature, some of the graphite (carbon) dissolves in the ferrite and can influence hardness. This depends upon the initial structure and the time-temperature relationship; even a 100% pearlitic matrix increases in carbon content at the austenitizing temperature if silicon content is low. Thus, it becomes a matter of controlling the time at temperature.

The methods of measuring hardenability are the same as those used for steel, which are described in Chapter 3.

Chapter 10

Case Hardening of Steel

There are two distinctly different approaches to case hardening, or the production of parts that have hard, wear-resistant surfaces, but with softer and/or tougher cores. One approach is to use a grade of steel that already contains sufficient carbon to provide the required hardness on heating and cooling; then, heating and quenching of only those portions that require hardening follow. The second method is to use a steel that is not normally capable of being hardened to a high degree, then alter the composition of the surface layers so that it either can be hardened or, in some instances, becomes hard during processing.

The first approach is discussed in Chapter 11 on flame and induction hardening. This chapter is confined to the discussion of the second approach—hardening processes that involve changes in surface composition.

Classification of Case Hardening

Because each general type of case hardening that involves composition changes incorporates a "family" of several types, precise classification becomes difficult. However, for most practical purposes, case hardening treatments can be broadly classified into four groups: carburizing, carbonitriding, nitriding, and nitrocarburizing. Of these processes, carburizing is by far the most widely used.

Carburizing Processes

In carburizing, austenitized ferrous metal is brought into contact with an environment of sufficient carbon potential to cause absorption of carbon at the surface and, by diffusion, to create a carbon concentration gradient between the surface and the interior of the metal. Two factors may control the rate of carburizing—the carbon-absorption reaction at the surface and diffusion of carbon into the metal.

Carburizing is done at elevated temperature, generally in the range of 1550 to 1750 °F (850 to 950 °C). However, temperatures as low as 1450 °F (790 °C) and as high as 2000 °F (950 °C) have been used. Although the carburizing rate can be greatly increased at temperatures above about 1750 °F (950 °C), most furnace equipment has an increasingly limited life, causing the carburizing temperature to be limited.

Carburizing may be done in a gaseous environment (gas carburizing), a liquid salt bath (liquid carburizing), or with all the surfaces of the workpiece covered with a solid carbonaceous compound (pack carburizing). Regardless of the process, the objective of carburizing is to start with a relatively low-carbon steel (0.20% C) and increase the carbon content in the surface layers, resulting in a high-carbon, hardenable steel on the outside with a gradually decreasing carbon content to the underlying layers. Process control to keep the desired carbon level is an important part of any carburizing process. Usually, although not always, a eutectoid carbon content (approximately 0.80%) is desired. Of the various carburizing processes, gas carburizing is the most widely used.

Gas Carburizing

In gas carburizing, the carbon source usually is introduced into an essentially noncarburizing or weak carburizing carrier gas. In general, gas carburizing is more effective than pack or liquid carburizing, and deeper and higher carbon content cases may be obtained more rapidly.

Gas carburizing may be more economical for mass production and can be mechanized to a greater extent than other carburizing processes. The economy is brought about because a specified case depth may be achieved more rapidly in gas carburizing. Also, gas carburizing does not involve the handling labor of packing and emptying the containers.

Carburizing Gases. Natural gas, which is largely methane (CH_4), is the most common source of carbon for gas carburizing. However, undiluted natural gas is much too rich to use directly so that only small quantitites are mixed with a noncarburizing gas (or near noncarburizing), known as the carrier gas.

The carrier gas is usually an externally produced gas such as endothermic gas (see section on furnace atmospheres in Chapter 4).

Gas carburizing also may be done using bulk nitrogen as the carrier gas. There are several different systems in use. In some of them, the nitrogen and carbon-source gas are carefully metered and mixed outside the furnace. While in other systems, the two gases are mixed inside the furnace chamber. Results obtained with the various types of carrier gases are essentially the same.

Liquid hydrocarbons are also extensively used as sources of carbon. These liquids are usually proprietary compounds that range in composition from pure hydrocarbons, such as dipentene or benzene, to oxygenated hydrocarbons, such as alcohols, glycols, or ketones. This process is sometimes called "drip carburizing" and should not be confused with liquid carburizing to be discussed later.

A liquid normally is fed in droplet form to a target plate in the furnace where it volatilizes almost instantaneously. The liquid dissociates thermally to provide a carburizing atmosphere. Forced fan circulation serves to distribute the atmosphere evenly throughout the furnace and also provides temperature uniformity. The liquid flow can be adjusted manually, although it can be a part of an automatically controlled system that provides almost any desired carbon potential.

Furnaces. In addition to vacuum furnaces, a wide variety of conventional furnaces (including pit, rotary, box, and continuous) are used for gas carburizing (see illustrations of atmosphere furnaces in Chapter 4. There are no specific limitations on furnace size. Selection of furnace type depends largely on workpiece shape and size, total production, and production flow, as for any heat treating process.

Characteristics of a Carburized Case. It is appropriate at this point to acquaint the reader with a few fundamentals regarding a gas carburized case. These generalizations also apply to cases produced in solid or liquid carbonaceous environments.

First, it must be understood that carburizing is a distinctly separate operation; that is, it is the process of diffusing carbon into the steel surface to make it hardenable and does not necessarily incorporate the hardening phase of the total operation. However, in most gas and liquid (salt bath) carburizing operations, the workpiece is quenched from an elevated temperature following the carburizing cycle. Therefore, in this specific section, it will be considered that the carburized work is also hardened in a continuous cycle.

A second important consideration is the nature of a carburized case, which is often poorly understood by design and process engineers. At a temperature of approximately 1700 °F (925 °C), a steel surface is extremely active, and if the carbon content of its environment is higher than that of the steel, the steel absorbs carbon to achieve equilibrium with the environment. However, if the carbon potential of the environment is lower than that of the steel, the steel loses carbon to its environment (decarburization). However, equilibrium conditions prevail only at the steel surface, and as distance from the surface increases, carbon concentration decreases gradually to the original carbon content of the steel. This is illustrated in Fig. 1, which relates carbon content (vertical axis) with distance from the surface (horizontal axis). For the conditions given in Fig. 1 for an 8620 H steel, the carbon content of the case decreases as distance from the surface increases. For example, the surface carbon is shown as 1.0%, but even at a depth of 0.020 in. (0.5 mm) the carbon concentration has decreased to about 0.80%. This clearly illustrates a very important fact—finishing operations (grinding or other) must be planned carefully, because too much stock removal in finishing removes the valuable part of the case. A general rule is to plan operations so that no more than 10% (per side) of the case is removed during finishing.

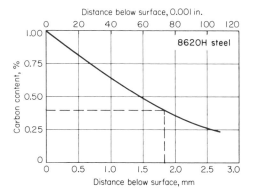

Fig. 1 Relationship of carbon concentration with distance from the surface. Source: *Carburizing and Carbonitriding*, American Society for Metals, p 21, 1977.

Total Case Versus Effective Case. Figure 1 also illustrates the principle of case depth. Total case (regardless of the carburizing method) refers to the very end point of the case—where the carbon content drops to the original level of the base metal (0.20% C in Fig. 1). In Fig. 1, this occurs at approximately 0.10 in. (2.5 mm). However, the "last part" of the carburized case is of little value and merely offers a measuring end point.

The term "effective case" is now used almost universally by industry. It is generally agreed that effective case is the point at which, after removal of the surface in small increments, hardness drops below 50 HRC.

In Fig. 1, the intersection of the dashed lines shows the effective case, which ends at approximately 0.075 in. (1.91 mm) and at the 0.40% C level. Depth of any carburized case is a function of time and temperature, as discussed below.

Effect of Temperature. The maximum rate at which carbon can be added to steel is limited by the rate of diffusion of carbon in austenite. This diffusion rate increases greatly with temperature; the rate of carbon addition at 1700 °F (925 °C) is about 40% greater than at 1600 °F (870 °C). For example, note Table 1; at 1600 °F (870 °C), a case depth of 0.025 in. (0.64 mm) is attained in 2 h, whereas a case depth of 0.035 in. (0.83 mm) is developed in the same length of time at 1700 °F (925 °C). It is mainly for this reason that the most common temperature for gas carburizing is 1700 °F (925 °C). This temperature permits a reasonably rapid carburizing rate without excessive deterioration of furnace equipment, particularly of heat-resistant alloys. Recently, there has been a trend toward raising the carburizing temperature to 1750 °F (955 °C) (or even 1800 °F or 980 °C), for certain deep-case requirements. For shallow-case carburizing in which case depth must be kept within a specified narrow range, lower temperatures are frequently used, because case depth can be more accurately controlled with the slower carburizing rates obtained with lower temperatures.

Data concerning several carburizing variables are presented in Fig. 2, but the effect of temperature is evident in interrelating the four sets of graphs that include case depth data for carburizing at 1600 °F (870 °C), 1650 °F (900 °C), 1700 °F (925 °C), and 1750 °F (955 °C), respectively.

Effect of Time. The depth of case achieved versus time is by no means a straight-line relationship, but it is more nearly like an inverse square relationship. Referring again to Table 1, it can be seen that the gain in case depth with increasing time at the carburizing temperature is a matter of rapidly diminishing returns. For example, at any of the three temperatures shown, to double the case depth obtained in 2 h, the time is approximately quadrupled. This fact is also borne out in the temperature—case depth—time data shown in Fig. 2. Note that as a matter of convenience, the time data are plotted on a logarithmic scale. Formulas are available for more precise calculations of case depth (see Ref 20).

Control of Carbon Concentration. The means for determining and controlling the carbon potential of the gas is quite complex and cannot be completely described here. However, several systems have been devised for this type of control so that all an operator needs to do is set the instrument for the desired surface carbon content; the gas mixture is controlled accordingly.

Because amount of water vapor in the furnace atmosphere is directly related to carbon potential, dew point has been the most commonly used method of controlling carbon potential. However, there are several other control systems including infrared analyzers, dew point analyzers, oxygen probes, and hot wire analyzers. Any of these systems may be utilized to control the carbon potential automatically within the pre-established limits.

Table 1 Values of case depth for various times and temperatures
Source: Metals Engineering Institute Course 10, Lesson 9, p 12, 1979.

Time, t, h	Case depth, in., after carburizing at		
	1600 °F	1650 °F	1700 °F
2	0.025	0.030	0.035
4	0.035	0.042	0.050
8	0.050	0.060	0.071
12	0.061	0.073	0.087
16	0.071	0.084	0.100
20	0.079	0.094	0.112
24	0.086	0.103	0.122
30	0.097	0.116	0.137
36	0.108	0.126	0.150

Vacuum Carburizing

Vacuum carburizing is essentially a gas carburizing process that is carried out at a pressure less than atmospheric (100 kPa or 760 torr). Carburizing temperatures may range from 1650 to 2000 °F (900 to 1100 °C), but usually are 1800 to 1925 °F (980 to 1050 °C).

The procedure involves pumping the furnace down to a soft vacuum then flowing the atmosphere gas into the chamber at carburizing temperature. The atmosphere gas usually consists solely of enriching gas; nitrogen also may be used as a carrier gas.

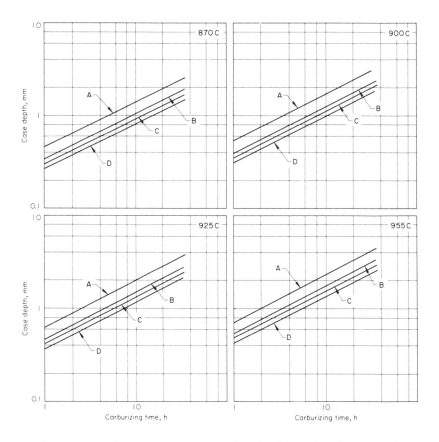

Fig. 2 Case depth as a function of carburizing time for normal carburizing (no diffusion cycle) of low-carbon and certain low-alloy steels. Curve A: Total case depth. Curve B: Effective case depth for surface carbon content of 1.1% to saturation. Curve C: Effective case depth for surface carbon content of 0.8 to 0.9%. Curve D: Effective case depth for surface carbon content of 0.7 to 0.8%. Source: *Carburizing and Carbonitriding*, American Society for Metals, p 68, 1977.

Generally, graphite heating elements and fixtures are used instead of metals in furnace construction. Therefore, higher carburizing temperatures may be used for vacuum carburizing than for other gas carburizing methods.

The principal advantage of the vacuum method is faster carburizing which is due to higher carburizing temperatures. Also advantageous is the fact that no atmosphere generator is required, and less carburizing gas is needed.

Liquid Carburizing

The term "liquid carburizing" should not be confused with drip carburizing. Liquid carburizing is a method of case hardening ferrous metal parts by holding them above their transformation temperature in a molten salt bath. The salt decomposes and releases carbon and sometimes nitrogen that diffuse into the work metal surfaces so that a high degree of hardness can be developed on quenching.

Many liquid carburizing baths contain cyanide, which introduces both carbon and nitrogen into the case. A new type of bath, which is enjoying some success, uses a special grade of carbon rather than cyanide as the source of carbon. This bath produces a case that contains only carbon as the hardening agent.

Cyanide-Type Baths. Light case and deep case are arbitrary terms that have been associated with liquid carburizing in baths containing cyanide. There is necessarily some overlapping of bath compositions for the two types of cases. In general, the two types are distinguished more by operating temperature than by bath composition. Hence, the terms low temperature and high temperature are preferred.

Table 2 Composition of cyanide-type liquid carburizing baths
Source: Carburizing and Carbonitriding, American Society for Metals, p 148, 1977.

Constituent	Composition of bath, %	
	Light case, low-temperature (1550 to 1650 °F)	Deep case, high-temperature (1650 to 1750 °F)
Sodium cyanide	10 to 23	6 to 16
Barium chloride	...	30 to 55
Salts of alkaline earth metals (a)	0 to 10	0 to 10
Potassium chloride	0 to 25	0 to 20
Sodium chloride	20 to 40	0 to 20
Sodium carbonate	40 max	30 max
Accelerators other than those involving compounds of alkaline earth metals (b)	0 to 5	0 to 2
Sodium cyanate	1.0 max	0.5 max
Density of molten salt	1760 kg/m³	2000 kg/m³

(a) Calcium and strontium chlorides have been employed. Calcium chloride is the more effective, but its hygroscopic nature has limited its use. (b) Among these accelerators are manganese dioxide, boron oxide, sodium fluoride, and silicon carbide.

Both low-temperature and high-temperature baths are supplied in different cyanide contents to satisfy individual requirements of carburizing activity (carbon potential) within the limitations of normal dragout and replenishment. In many instances, compatible companion compositions are available for starting the bath or for bath makeup, and for the regeneration or maintenance of carburizing activity.

Low-temperature cyanide-type baths (light case) are those usually operated in the temperature range from about 1550 to 1650 °F (845 to 900 °C), although for certain specific effects, this temperature range is sometimes extended to cover operation from about 1450 to 1700 °F (790 to 925 °C). Low-temperature baths are best suited to the formation of cases 0.003 to 0.030 in. (0.075 to 0.75 mm) deep. They are accelerated cyanogen baths containing various combinations and amounts of the constituents listed in Table 2.

High-temperature cyanide-type baths (deep-case) are usually operated in the temperature range of about 1650 to 1750 °F (900 to 955 °C). This range may be extended somewhat, but at lower temperatures, the carbon penetration rate becomes undesirably slow. Also, at temperatures higher than about 1750 °F (955 °C), deterioration of the bath and equipment is markedly accelerated.

The most important use of high-temperature baths is for producing case depths of about 0.020 to 0.080 in. (0.5 to 2 mm), although deeper cases are possible. The baths consist of cyanide and a major proportion of barium chloride, with or without supplemental acceleration from other salts of the alkaline earth metals.

Noncyanide Liquid Carburizing. A bath that contains a special grade of carbon as the source of carbon has increased in commercial use. In this bath, carbon particles are dispersed in the molten salt by mechanical agitation, which is achieved with one or more simple propeller stirrers.

Operating temperatures for this type of bath are generally higher than for cyanide-type baths. A range of about 1650 to 1750 °F (900 to 955 °C) is most commonly employed. Temperatures below about 1600 °F (870 °C) are not recommended and may even lead to decarburization of the steel. The case depths and carbon gradients produced are in the same range as that for high-temperature cyanide-type baths.

Control of Case Depth and Carbon Concentration. In liquid carburizing, depth of case depends on time and temperature, as has been discussed for gas carburizing. This is indicated by Fig. 3, which shows the effects of three different carburizing temperatures for 1020 steel for time periods of 2 to 40 h.

Carbon concentration is controlled principally by control of the salt composition. The carbon and/or nitrogen that diffuses into the steel comes from decomposition of the salt. Therefore, any salt bath requires a great deal of attention in terms of baling out the sludge and replenishing with fresh salt—usually every 8-h shift, depending on the amount of production and, therefore, the amount of dragout.

Furnaces. In general, all of the salt bath furnaces discussed and illustrated in Chapter 4 are suitable for salt bath carburizing.

Advantages and Limitations of Liquid Carburizing. Two specific advantages of liquid carburizing are realized. Selective carburizing can be achieved in many applications without stop-off procedures. For example, to carburize and harden only one end of a long shaft-like member, the workpiece, by suitable fixturing, can be held so that only the end is immersed in the bath. This is not possible with other carburizing methods. A second advantage is the flexibility of liquid carburizing in simultaneously processing a variety of workpieces. They can vary in size and shape as well as case depth requirements for parts.

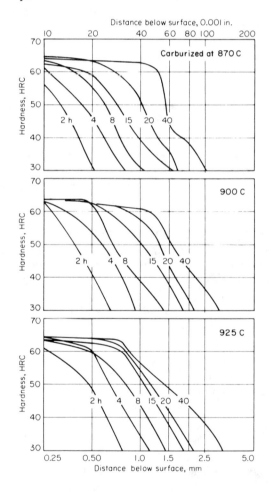

Fig. 3 Effects of time and temperature for liquid carburizing of 1020 steel. Source: *Carburizing and Carbonitriding*, American Society for Metals, p 152, 1977.

A disadvantage of liquid carburizing is the requirement of washing after quenching. Another is that the salt adhering to the hot workpieces contaminates quenching mediums. Because of the problems associated with salt removal, liquid carburizing is not recommended for parts containing small holes, threads or recessed areas that are difficult to clean.

Pack Carburizing

Pack carburizing is a process in which carbon monoxide derived from a solid compound decomposes at the metal surface into nascent carbon and carbon dioxide. The nascent carbon is then absorbed into the metal. The carbon dioxide resulting from this decomposition immediately reacts with carbonaceous material present in the solid carburizing compound to produce fresh carbon monoxide. This reaction is enhanced by energizers or catalysts, such as barium carbonate ($BaCO_3$ and sodium carbonate (Na_2CO_3), that are present in the carburizing compound. These substances react with carbon to form additional carbon monoxide and an oxide of the energizing compound. The latter in turn reacts in part with carbon dioxide to re-form carbonate. Thus, in a closed system, the energizer is continuously being used and re-formed. Carburizing continues as long as enough carbon is present to react with the excess of carbon dioxide.

For the most part, pack carburizing has been replaced by other methods. However, there are special applications where pack carburizing is desired.

Furnaces. Pack carburizing is usually done in box batch-type furnaces. No prepared atmosphere is required. Typical furnaces used for this process are discussed in Chapter 4.

Processing. Parts to be processed are placed in steel or heat-resistant alloy boxes surrounded by a generous amount of the solid carburizing compound, usually purchased as a proprietary mixture. The boxes are then sealed with fire clay and heated to the carburizing temperature for the necessary time. Carburizing rates are generally lower for pack carburizing, but otherwise the same variables prevail. There are occasions where, at the end of the carburizing cycle, parts are removed from the boxes and directly quenched for hardening. As a rule, however, parts are slow cooled in the boxes, then reheated for hardening.

Carbon Potential and Gradient. The carbon potential of the atmosphere generated by the carburizing compound, as well as the carbon content obtained at the work's surface, increases directly with an increase in the carbon monoxide to carbon dioxide ratio. Thus, more carbon is made available at the work surface by the use of energizers and carburizing materials that promote carbon monoxide formation.

The carbon concentration gradient of carburized parts is influenced principally by carbon potential, carburizing temperature and time, and the chemical composition of the steel.

Pack carburizing is normally performed in the temperature range of 1500 to 1700 °F (815 to 955 °C), although carburizing temperatures as high as 2000 °F (1095 °C) have been used.

The carburizing rate is more rapid at the start of the cycle and gradually diminishes as the cycle is extended as shown for gas carburizing.

Process Limitations. Principal disadvantages of pack carburizing as opposed to other carburizing includes the fact that pack carburizing is grossly inefficient in terms of energy consumption because of all the material (boxes and compound) that must be heated, which usually amounts to the weight of the workpieces or more. Also, the difficulty of direct quenching, thus consuming more energy for reheating, is a disadvantage.

Carbonitriding

Carbonitriding is a modified gas carburizing process, rather than a form of nitriding. The modification consists of introducing ammonia (commonly about 10%) into the gas carburizing atmosphere in order to add nitrogen to the carburized case as it is being produced. Ammonia in the atmosphere dissociates to form nascent nitrogen at the work surface, and the nascent nitrogen diffuses into the steel simultaneously with carbon. Typically, carbonitriding is carried out at a lower temperature and for a shorter time than gas carburizing, in order to obtain a case shallower than is usual in production carburizing. Lower temperatures are possible in carbonitriding primarily because nitrogen is a powerful austenitizer; thus, the transformation temperature for any given steel is lowered.

In its effects on steel, carbonitriding is similar to liquid cyaniding. Because of problems in disposing of cyanide-bearing wastes, and washing problems, carbonitriding is often preferred over liquid cyaniding. In terms of case characteristics, carbonitriding differs from carburizing and nitriding in that carburized cases normally do not contain nitrogen and nitrided cases contain no added carbon, whereas carbonitrided cases contain both.

Applications. Carbonitriding is used mainly to impart a hard, wear-resistant but relatively thin case to a large variety of hardware items (mostly small sized) on a mass-production basis. Case depths usually range in thickness from 0.003 to 0.030 in. (0.075 to 0.75 mm); but largely fall near the lower end of this range. Figure 4 shows typical hardness-case depth relationships for the two steels.

A carbonitrided case has higher hardenability than a carburized case; consequently, by carbonitriding and quenching with either a carbon or low-alloy steel, a hardened case can be produced at less expense within the case-depth range indicated. Because of the higher hardenability of carbonitrided cases, full hardness can often be obtained by oil quenching instead of quenching in water or brine as might be required for a conventional carburized case. Thus, less distortion of the workpiece results.

Steels for carbonitriding are generally of the lower carbon variety (0.30% C max for plain carbon or alloy), although sometimes steels having a somewhat higher carbon content (0.30 to 0.35% C) are case hardened by carbonitriding.

Furnaces that are suited to gas carburizing are generally suitable for carbonitriding. However, because carbonitriding is usually a mass-production process, some form of conveyor furnace is usually used (see Chapter 4).

Any atmosphere composition that can be used effectively for gas carburizing can be used for carbonitriding, with an ammonia addition. For more details on this process, see Ref 19 and 20.

Conventional Gas Nitriding

Gas nitriding is a case hardening process in which nitrogen is introduced into the surface of a ferrous alloy by holding the metal at a suitable temperature (below Ac_1 for ferritic steels) in contact with a nitrogenous gas, usually ammonia. At elevated temperature, the ammonia dissociates into its components according to the reaction:

$$2NH_3 \rightarrow 2N + 3H_2$$

The nitrogen, which is very active at the moment of decomposition of the ammonia gas, combines with the alloying elements in the steel to form nitrides. These nitrides form at the steel surface as a fine dispersion and impart extremely high hardness to the steel surface without need for quenching.

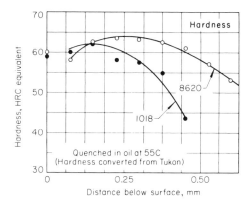

Fig. 4 Hardness-depth relationships for carbonitriding of one plain carbon and one alloy steel. Source: *Carburizing and Carbonitriding*, American Society for Metals, p 128, 1977.

Steels for Gas Nitriding. Because aluminium is the strongest nitride former of the common alloying elements, aluminum-containing steels (0.85 to 1.50% Al) yield the best nitriding results in terms of total alloy content. Chromium-containing steels can approximate these results if their chromium content is high enough (approximately 5% Cr). Plain carbon steels are not suited to gas nitriding, because they form an extremely hard, brittle case that spalls readily. Compositions of some typical aluminum-bearing steels are shown in Table 3.

Other steels that can be surface hardened by gas nitriding include medium-carbon, chromium-containing alloy steels; hot work tool steels such as H11, H12, and H13; and most stainless steels. In conventional gas nitriding, best results can be obtained only by hardening and tempering the steels before nitriding (see Table 3).

Furnaces. As a rule, gas nitriding is done in batch-type furnaces such as a pit or bell-type furnace (see Chapter 4). A major requirement is that the furnace can be tightly sealed to prevent infiltration of air.

Processing. Either a single- or a double-stage process may be used. In the single-stage process, a temperature range of about 925 to 975 °F (495 to 525 °C) is used, and the dissociation rate ranges from 15 to 30%.

The first stage of the double-stage process, is, except for time, a duplication of the single-stage process. The second stage may proceed at the nitriding temperature employed for the first stage, or the temperature may be increased from 1025 to 1050 °F (550 to 565 °C), but at either temperature, the rate of dissociation in the second stage is increased from 65 to 85% (preferably 80 to 85%). Generally, an ammonia dissociator is necessary to obtain the required higher second-stage dissociation. In all cases, the furnace must be purged with ammonia or nitrogen so that the atmosphere contains no more than 5% of air before the nitriding cycle is started. Cycles may range from 10 to 50 h or more.

Case Characteristics. Nitrided cases are much harder (when correctly evaluated) than carburized cases, but they are thin and have a hardness gradient not unlike a carburized case. Nitrided cases also have greater resistance to softening from heat compared with carburized cases.

Salt Bath Nitriding

Steels that can be nitrided in a gaseous atmosphere may also be nitrided in molten salt (liquid nitriding) at essentially the same temperature—950 to 1050 °F (510 to 565 °C). As in liquid carburizing and cyaniding, the case hardening medium is molten cyanide. Unlike liquid carburizing and cyaniding, however, liquid nitriding is done below, instead of above, the transformation temperature range of the steel being treated.

Furnaces for salt bath nitriding are the same as those used for liquid carburizing (see Chapter 4).

Types of Salts. A typical commercial bath for liquid nitriding is composed of a mixture of sodium and potassium salts. The sodium salts, which comprise 60 to 70 wt% of the total mixture, consist of 96.5% NaCN, 2.5% Na_2CO_3, and

Table 3 Nominal compositions and preliminary heat treating cycles for aluminum containing low-alloy steels commonly gas nitrided

Source: Metals Engineering Institute Course 10, Lesson 9, American Society for Metals, p 28, 1979.

Steel		Nitralloy	Composition, %								Austenitizing temperature, °F (a)	Tempering temperature, °F (a)
SAE	AMS		C	Mn	Si	Cr	Ni	Mo	Al	Se		
...	...	G	0.35	0.55	0.30	1.2	...	0.20	1.0	...	1750	1050-1300
7140	6470	135M	0.42	0.55	0.30	1.6	...	0.38	1.0	...	1750	1050-1300
...	6475	N	0.24	0.55	0.30	1.15	3.5	0.25	1.0	...	1650	1200-1250
...	...	EZ	0.35	0.80	0.30	1.25	...	0.20	1.0	0.20	1750	1050-1300

(a) Sections up to 2 in. in diameter, quenched in oil; larger sections may be water quenched.

0.5% NaCNO. The potassium salts, 30 to 40 wt% of the mixture, consist of 96% KCN, 0.6% K_2CO_3, 0.75% KCNO, and 0.5% KCl. The operating temperature of this salt bath is 1050 °F (565 °C). Control of bath composition is critical.

Proprietary Salt Bath Processes. Aerated salt bath nitriding is a proprietary process that employs cyanide salts. In this process, measured amounts of air are pumped through the molten bath. The introduction of air provides agitation and stimulates chemical activity. The cyanide content of this bath, calculated as NaCN, is preferably maintained at about 50 to 60% of the total bath content, and the cyanate at 32 to 38%. The potassium content of the fused bath, calculated as elemental potassium, is between 10 and 30%, preferably about 18%. The potassium may be present as the cyanate or the cyanide, or both. The remainder of the bath is sodium carbonate.

This process produces a nitrogen-diffused case 0.012 in. (0.3 mm) deep on plain carbon or low-alloy steels in a 1½-h cycle. The surface layer (0.0002 to 0.0005 in., or 0.005 to 0.013 mm deep) of the case is composed of epsilon nitride (Fe_3N) and a nitrogen-bearing Fe_3C. Parts that are nitrided by the aerated bath process are machined to final dimensions before being nitrided.

Another proprietary aerated salt bath nitriding process is capable of producing results similar to the process discussed above. The advantage of this process is that the salts do not contain cyanide.

Gaseous Nitrocarburizing

Nitrocarburizing is a relatively new term, which includes several proprietary names. In general, the results are similar to those obtained with aerated salt bath nitriding (sometimes called salt bath nitrocarburizing). A natural advantage of the gas process is the elimination of the need for salt removal.

Ferritic nitrocarburizing treatments involve diffusion of both nitrogen and carbon into the surfaces of ferrous metals at temperatures below 1250 °F (675 °C), thus distinguishing it from carbonitriding because nitrocarburizing is done with steels in their ferritic phase.

The primary object of such treatments is usually to improve antiscuffing characteristics of ferrous engineering components by providing the surface with a compound layer—really, a surface zone—exhibiting good wear/friction-resistant properties. In addition, fatigue characteristics can be considerably improved, particularly when nitrogen is retained in solid solution in the "diffusion zone" beneath the compound layer. This retention normally is achieved by quenching in oil or water from the treatment temperature. Corrosion resistance provided by the compound zone is an important secondary benefit.

Gaseous nitrocarburizing commonly employs sealed-quench batch furnaces of the same design used for carburizing and carbonitriding (see

Chapter 4). Furnace operating temperatures are low enough to maintain steels in the ferritic condition. The atmosphere employed consists of ammonia diluted with a carrier gas. In one process, the atmosphere is formed from equal amounts of ammonia and endothermic gas. In another process, a typical atmosphere consists of 35% ammonia and 65% refined exothermic gas (nominally 97% nitrogen), which may be enriched with a hydrocarbon gas. High-purity nitrogen is used as a dilutent in one of these processes. Gaseous nitrocarburizing is performed near 1060 °F (570 °C), a temperature below the austenite range for the iron-nitrogen system. Treatment times generally range from 1 to 5 h.

Chapter 11

Flame and Induction Hardening

Case hardening of steel (the subject of Chapter 10) dealt exclusively with methods that involved changes in surface composition. Methods discussed in this chapter produce hard surfaces (commonly called cases) on ferrous metal parts (steels and cast irons) with no change in composition. While flame and induction hardening are the principal methods used, there are two other methods of producing parts with hard surfaces without a change in composition.

Hardened Zones in Low-Hardenability Steels

The use of low-hardenability steels is one of the oldest methods of producing a part with a hard surface and a relatively soft core without altering composition. Although the part is heated throughout the section, a low-hardenability steel transforms from the austenitic state to softer products such as ferrite or pearlite so rapidly that only the outer layer is cooled quickly enough to form martensite. Figure 1 shows that only a narrow band of full hardness is obtained after heating and quenching. Because of the lesser mass involved, the depth of the hardened zone is greater on the smallest diameter section (see also Chapter 8).

Shell Hardening in Molten Metal or Salt

Another method of hardening only the surface layers of a workpiece, before involving flame and induction methods, is by a technique commonly known as shell hardening. The workpiece is immersed in a high-conductivity heating medium such as molten lead or a salt bath, but only long enough to completely heat the outer layers. This is followed by quenching. By this technique, the core portion is never heated sufficiently to form austenite. This technique is limited in use, largely by part design. Obviously, the part must have some appreciable minimum thickness or it would heat completely through. Cold working dies are typical examples of this application to surface hardening.

Flame Hardening

One of the oldest methods of surface, or selective area, hardening is by means of heating the area with an oxyacetylene flame, which supplies a very intense and concentrated heat.

Using this method, the hardened layer may be varied from a very thin skin to depths of 1/4 in. (6 mm). The depth of the hardened zone is controlled by the amount of time the torch is allowed to heat the steel. With short heating times, only a thin skin is made austenitic and hardened. With longer heating times, the heat penetrates to a greater depth and results in deeper hardening. All sorts of sections and sizes, as well as any accessible portion of a workpiece, can be hardened by this procedure. Likewise, quenching procedures may vary and are not necessarily the same as for the same steel heated and quenched from a furnace. This is because the quickly heated area receives a "mass quenching effect" from the mass of underlying cold metal. In many instances where the heated zone is shallow and the mass is relatively large, liquid is not required to develop full hardness. Air blast cooling may achieve the same effects.

Almost any new flame hardening operation requires developmental work (cut and try) to arrive at the optimum procedure for a specific workpiece. Past experience with similar workpieces is always helpful in developing specific procedures.

Material for flame hardening can be any ferrous metal (steel or cast iron) that has the potential for being directly hardened by any other method.

Equipment choices for flame hardening range from a hand-held torch to an elaborate mechanized setup, depending mainly on the number of identical parts to be processed.

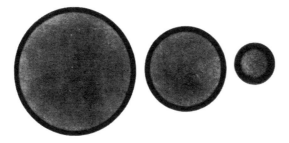

Fig. 1 Cross section of three sizes of water-hardening tool steel (WI) after heating to 1475 °F (800 °C) and quenching in brine. Black rings indicate hardened zones (cases) (65 HRC). Cores range from 38 to 43 HRC. Source: Metals Engineering Institute Course 10, Lesson 9, American Society for Metals, p 2, 1979.

Methods of Flame Hardening. Principal procedures are spot or stationary, progressive, spinning, and combination progressive-spinning modes of operation. Selection of the most appropriate system depends on shape, size, and composition of the workpiece; the area to be surface hardened; required depth of the hardened zone; and the number of pieces to be hardened.

Figure 2(a) illustrates the spot or stationary method, consisting of heating specific areas with a suitable flame head and subsequently quenching. Both torch and workpiece are kept stationary, and only the area immediately under the torch is heated and hardened. This is the simplest of the four systems because mechanical equipment is not required (except perhaps a fixture and timing device to ensure uniform processing of each workpiece). If desired, however, the operation can be automated.

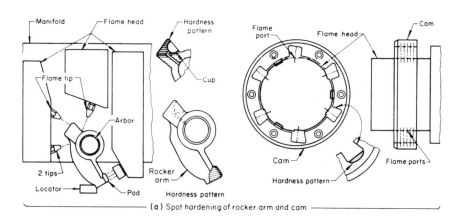

(a) Spot hardening of rocker arm and cam

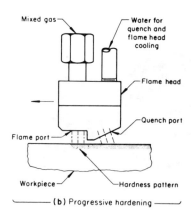

(b) Progressive hardening

Fig. 2 Methods of flame heating. (a) Spot (stationary) heating a rocker arm and the internal lobes of a cam, quench not shown. (b) Progressive method. Source: Metals Engineering Institute Course 10, Lesson 9, American Society for Metals, p 3, 1979.

The progressive method is used to harden large areas that are beyond the scope of spot hardening. As illustrated in Fig. 2(b), the heating torch and integral quenching head travel along the face of the object to be hardened. The torch and quench head also may be kept stationary, and the work moved along under it. Speeds of about 3 to 12 in./min (75 to 300 mm/min) are commonly employed. Thus, the work surface is progressively heated and hardened as either the torch or the work moves along.

This method of flame hardening can be applied to various workpiece shapes and sizes, but is used most often for hardening of long flat areas such as machine tool ways.

A distinct disadvantage of the method illustrated in Fig. 2(b) for flame hardening of rounds is that it is virtually impossible to avoid formation of a narrow zone that is slightly softer than the rest of the hardened skin, which is left at the starting-stopping point. In high-carbon steels, cracking also may be encountered at this zone when the flame overlaps the previously hardened area. For these reasons, it is usually preferable to harden cylindrical workpieces by spinning, as shown in Fig. 3 or 4.

Equipment for hardening by the progressive method commonly consists of one or more flame heads and a quenching means mounted on a carriage that runs on a track at a regulated speed.

The spinning method is commonly employed for hardening small rounds. The torch usually is held stationary and the work rotated at speeds of up to 100 rpm. The quenching water is turned off while the piece is being heated. The work is rotated until the surface reaches the desired temperature or until the heat penetrates to the desired depth. At this time the flames are extinguished and the quenching water turned on, or the part is dropped into a quenching tank. Sometimes, multiple torches are used to ensure rapid and uniform heating. Three methods of spin hardening are illustrated in Fig. 3. In two of the methods, the workpiece is rotated. However, in the third application (at the right), rotation of the flame head was more practical.

A combination of the progressive and spinning methods is illustrated in Fig. 4. Frequently used for hardening the surfaces of long cylinders, this method consists of moving a heating ring along the length of the cylinder at speeds of about 3 to 12 in./min (75 to 300 mm/min), followed by a quenching ring. The progressive spinning method differs from the simple progressive method in that the work is spun while it is being heated. The rapidity of revolution depends on the size of the piece being hardened.

Multiple heating heads may be employed to ensure proper heating. In this method, as in the others, the depth of hardening depends on the time in the heating flame.

Quenching. In Fig. 2, 3, and 4, note that quenching facilities are "built in", which usually is the procedure for relatively small workpieces heat treated on a production basis. Larger workpieces may not require quenching.

In almost all instances, water is used for spray quenching following flame heating, for two reasons: (1) spraying of oil presents a fire hazard and (2) the

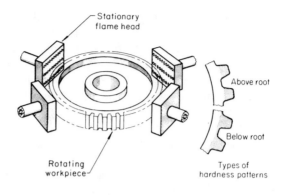

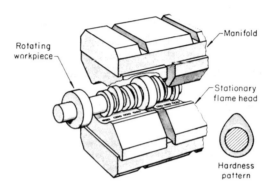

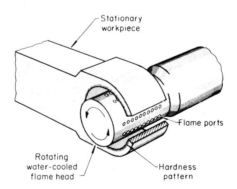

Fig. 3 Spinning methods of flame heating, in which (left, center) the part rotates and (right) flame head rotates. Quench is not shown. Source: Metals Engineering Institute Course 10, Lesson 9, American Society for Metals, p 4, 1979.

amount of water used can be closely controlled so that the workpiece does not come from the quench in a "stone cold" condition. As mentioned in Chapter 2, drastic quenching is required only to exceed the critical cooling rate down to the lower temperature range wherein cooling rate is no longer critical. The practical results are that steels (or cast irons) that might crack when water quenched by immersion can be safely hardened by control of the water in a spray quench.

Tempering. Even though the workpiece is still quite warm when removed from a flame heating and quenching operation, there still is a need for tempering. Regardless of the system used to flame harden a steel workpiece, the work must be tempered after the hardening operation. Tempering relieves the stress set up in hardening and imparts some degree of toughness to the hard case.

Induction Hardening

Electromagnetic induction is one method of heating a metal part. It can be used for normalizing, some annealing processes, heating for hardening, tempering, and heating of nonferrous metals. This method of heating can also be used for brazing.

The use of induction heating, as dealt with in this chapter, is confined principally to localized heating for subsequent hardening (usually surface hardening).

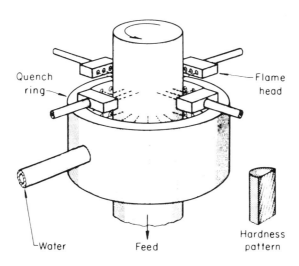

Fig. 4 Combination progressive-spinning method of flame hardening. Source: Metals Engineering Institute, Course 10, Lesson 9, American Society for Metals, p 5, 1979.

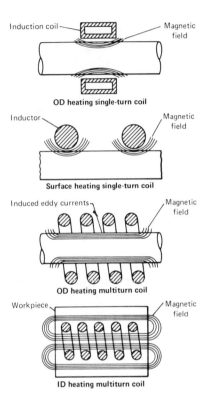

Fig. 5 Magnetic fields and induced currents produced by various induction coils. Source: *Metals Handbook*, Vol 4, 9th ed., American Society for Metals, p 451, 1981.

The metallurgical principles and advantages of induction heating, compared with furnace heating, generally are the same as those for flame hardening, described earlier in this chapter. Both of these methods are characterized by very rapid heating and the capability of permitting extremely close control of the quenching phase of the operation.

Principles of Induction Heating. Any electrical conductor can be heated by electromagnetic induction. As alternating current flows through the inductor, or work coil, a highly concentrated, rapidly alternating magnetic field is established within the coil. The strength of this field depends primarily on the magnitude of the current flowing in the coil. The magnetic field thus established induces an electric potential in the part to be heated, and because the part represents a closed circuit, the induced voltage causes the flow of current. The resistance of the part to the flow of the induced current causes heating.

The depth of current penetration depends upon workpiece permeability, resistivity, and the alternating current frequency. Because the first two factors vary comparatively little, the greatest variable is frequency. Depth of current penetration decreases as frequency increases. High-frequency current generally is used when shallow heating (thin case) is desired; intermediate and low frequencies are used in applications requiring deeper heating.

Most induction surface-hardening applications require comparatively high power densities and short heating cycles to restrict heating to the surface area. The principal metallurgical advantages that may be obtained by surface hardening with induction are the same as for flame hardening.

The pattern of heating obtained by induction is determined by the (1) shape of the induction coil producing the magnetic field, (2) number of turns in the coil, (3) operating frequency, (4) alternating current power input, and (5) nature of the workpiece. Four examples of magnetic fields and induced currents produced by induction coils are shown in Fig. 5.

Fig. 6 Induction hardening of a gear. The heating coils are surrounded by the spray quenching head. A thin surface layer of the gear will be heated, power turned off and the spray turned on, resulting in a thin hard case. Source: Metals Engineering Institute Course 10, Lesson 9, American Society for Metals, p 6, 1979.

The rate of heating obtained with induction coils depends on the strength of the magnetic field to which the part is exposed. In the workpiece, this becomes a function of the induced currents and of the resistance to their flow.

The paragraphs above describe what happens when steel is heated by induction, but do not explain why heat is developed in the workpiece. It is generally accepted that heat is developed and conducted to the interior in the following three ways.

One way is to place magnetic materials such as steel in a magnetic field, where the molecules tend to align themselves with the polarity of the field. If the current is reversed a great number of times per second, as in high frequency alternating current, the molecules also tend to change their alignment a similar number of times. As a result, molecular friction is set up and heat is generated. When the steel is heated and becomes austenitic, it also becomes nonmagnetic. The heat generated by this effect then becomes negligible.

Another method of developing and conducting heat to the interior is by inducing current into a steel workpiece. The induced current has a tendency to swirl in much the same fashion as a pool of water swirls when stirred with a stick. The swirling effect of the current is known as eddy current, and it induces heat. The eddy current effect probably is the major source of heat, particularly when the steel has become nonmagnetic.

In a third method, heat is generated in the surface layers and carried to the interior by simple conduction. After the steel has been heated to the proper austenitizing temperature and required depth, quenching is accomplished by introducing a water spray or other suitable quenching fluid through the spaces between the inductor coil, or by dropping the workpiece from its heating position in the coil into a quenching tank. The depth of hardness penetration is controlled by the power input, the frequency used, and time in the inductor. Because the heating effect is very rapid, hardening only a thin skin without quenching is possible. In such instances, the thin hot skin is cooled by the large cold mass lying under it. The piece is then described as self-quenched.

Table 1 Power sources, frequencies, efficiency and power ratings for induction heating equipment
Source: Metals Engineering Institute Course 10, Lesson 9, American Society for Metals, p 7, 1979.

Type	Frequency, kHz	Efficiency, %	Power rating, kW
Vacuum tube oscillators	200-450	50-60	5-600
Motor generators	1, 3, 10	75-80	7.5-500
Frequency multipliers	180 and 540 Hz	90-95	100-1000
Frequency inverters	0.5, 1, 3, 10	85-95	50-1500

An induction heating operation in its simplest form is shown in Fig. 6. A small gear is manually placed within a multiple-turn coil (the inductor). The coil is made from copper tubing to permit water cooling when in operation. The outer ring is the quench ring, which spray quenches the gear in a timed operation at the end of the heating cycle.

Equipment for Induction Hardening

The basic equipment for induction hardening consists of a power supply, controls, a means of matching the power supply with the load, an inductor, quenching equipment, and a system for holding and positioning the workpieces. The latter may range in degree of sophistication from a simple locator (as in Fig. 6) to a completely automated positioning system.

Fig. 7 An inductor with separate internal chambers for flow of quenchant and cooling water. Source: Metals Engineering Institute Course 10, Lesson 9, American Society for Metals, p 7, 1979.

Power Supplies. The four most widely used power supplies are vacuum tube oscillators, motor generators, frequency multipliers, and frequency inverters. Spark-gap oscillators and mercury converters are no longer manufactured and generally are considered obsolete, although some units of each type may still be in service.

The vacuum tube oscillator changes the power line voltage to 15 000 to 20 000 V and rectifies it to direct current. Then, by using a circuitry involving one or more vacuum tubes, capacitors, and coils, it generates high-voltage, high-frequency power at 200 to 450 kHz. Operating frequency is determined by workpiece size, inductor design, and load matching.

The motor generator consists of a conventional three-phase induction motor, which drives a high-frequency, single-phase alternator. The most common frequencies are 1 to 10 kHz. Frequency multipliers essentially are special transformers that triple the line frequency to single-phase 180 Hz. This may be tripled again to 540 Hz.

Frequency inverters use diodes and silicon controlled reactors. For the lower frequency range (up to 10 kHz), the inverter is gradually becoming the main source of high-frequency power for two principal reasons: (a) equipment cost is lower than for motor-generator installations in terms of kW delivered and (2) the inverter is more efficient because it does not use power when a workpiece is not being heated; the motor generator uses a significant amount of power when not heating a workpiece. Frequencies and efficiency and power ratings for the four principal types of induction heating equipment are given in Table 1.

Inductors. For the most efficient heating, the work should be surrounded by the coil with a minimum air gap between the inductor and the work. An inductor that meets these requirements is shown in Fig. 7. This single-turn inductor contains separate internal chambers—one for cooling water and one for the quenchant. With suitable auxiliary equipment, this general design can be used for a variety of applications including progressive surface hardening of long bar-like products.

Despite the fact that greatest efficiency is attained when the inductor surrounds the work, this is not always feasible. For this reason, there are many other inductor designs including single- and multiple-turn types. There are even specially designed inductors for hardening internal surfaces such as cylinder bores. However, these generally are less efficient than the type illustrated in Fig. 7. A variety of coils and resulting heat patterns are shown in Fig. 8, which illustrates some of the attainable objectives with induction heating, including the heating of specific areas on flat surfaces (Fig. 8e).

Selection of Frequency

The application generally dictates frequency requirements, although some broad overlapping exists. In many instances, essentially the same results can be achieved with more than one frequency by using different power densities and heating times. For example, a frequency of 10 kHz at low power and long heating time could duplicate the heating of 3 kHz at higher power and shorter heating time; the same is true for 400 and 10 kHz.

In establishing processing cycles for surface hardening, the minimum depths that are considered practical for 3, 10, and 450 kHz are given in Table 2. Producing shallower depths than are required is expensive because it requires higher frequency and/or higher power—both of which increase the original equipment and operating costs.

For most surface hardening applications, 10 or 3 kHz is satisfactory. Ten kHz provides a suitable contoured hardness pattern on cams and coarse pitch gears. Finer pitch gears require 450 kHz to prevent through-hardening. But in some instances, because the roots are not heated effectively at the higher frequency, 10 kHz is used in a dual frequency approach. The valleys are preheated with 10 kHz; 450 kHz is used for the tips. In all such applications, time must be kept short to limit heat flow by conduction.

Process Development. Not unlike flame hardening, development of optimum heating and cooling cycles for induction hardening of a specific

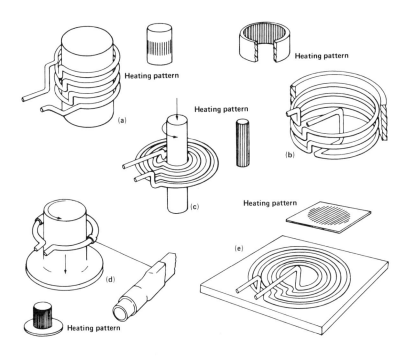

Fig. 8 Typical work coils for induction heating. Source: *Metals Handbook*, Vol 4, 9th ed., American Society for Metals, p 459, 1981.

Table 2 Minimum hardness depths for production work in surface hardening
Source: Metals Engineering Institute Course 10, Lesson 9, American Society for Metals, p 8, 1979.

Frequency, Hz	Hardness depth (a), in.	Penetration of electrical energy (b), in.
3 000	0.060	0.035
10 000	0.040	0.020
450 000	0.020	0.003

(a) Approximate practical minimum depth of hardness in inches. (b) Approximate theoretical depth of penetration of electrical energy in inches

workpiece is usually the result of some experimentation. For information that may be helpful in developing a new setup, see Ref 10. Past experience with similar workpieces usually is the most helpful tool in establishing new cycles of heating and cooling.

Chapter 12

Heat Treating
of Nonferrous Alloys

The reader should understand at the outset that this chapter is intended to present only an overview of the heat treating of nonferrous alloys. First, a brief discussion of the effects of cold work and annealing on nonferrous alloys is presented. This is followed by discussion of the mechanisms involved in the more common heat treating procedures used for hardening or strengthening—solution treating and aging. No attempt is made in this treatise to cover the more complex procedures required for duplex structures which characterize some nonferrous alloys, notably titanium and copper alloys. For more complete information on the heat treating of nonferrous alloys and the properties that may be obtained, see Ref 10 and 21.

Work Hardening

Most metals and metal alloys (ferrous and nonferrous) respond to hardening from cold work. There is, however, a wide variation among the many metal and metal alloy combinations in their rate of work hardening by successive cold working operations.

Basically, work hardening occurs by converting the original grain shape into deformed, elongated grains by rolling or other cold working procedures. This grain shape change is illustrated schematically in Fig. 1. In this case, a copper-zinc alloy is subjected to 60% reduction by cold rolling. The change in grain shape is evident. In practice, this much reduction in one pass would be rare, but the end result is essentially the same regardless of the number of passes.

In this example, hardness of the alloy was originally 78 HRB, but changed to 131 HRB after cold reduction. This amount of hardness increase is typical of many copper alloys, most of which are very high in response to work hardening.

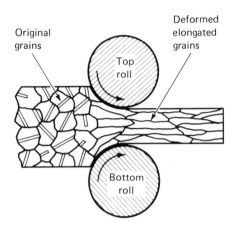

Fig. 1 Schematic presentation of cold rolling a copper-zinc alloy (60% reduction). Hardness was 78 HRB before reduction, which increased to 131 HRB after reduction. Source: *Heat Treatment, Structure and Properties of Nonferrous Alloys*, p 22, 1982.

Annealing of Nonferrous Metals and Alloys

Most nonferrous alloys that have become hardened by cold work can be essentially restored to their original grain structure by annealing, during which process recrystallization occurs. This process is closely related to process annealing of low-carbon steels, which includes a certain amount of resulting grain growth that can be controlled to a great extent by annealing time and temperature.

A typical example of the effects of annealing temperature on hardness (diamond pyramid or Vickers) of pure copper and a copper-zinc alloy is presented in Fig. 2. Several facts are evident:

- The copper-zinc alloy work hardens to a higher degree than pure copper
- The effects of annealing are evident at a much lower temperature for pure copper compared to its companion alloy
- Time at temperature has far less effect than temperature
- Hardness of the pure metal and the alloy are nearly equal when near full annealing is accomplished at approximately 1110 °F (600 °C).

Figure 2 is presented as a simple example of recrystallization annealing. Each of the many nonferrous metals and metal alloy combinations has unique annealing characteristics which produce a unique graph of data. In many instances, adjustment may be required in the annealing time and/or temperature to attain desired grain size. Although there are many sources for data that apply to specific alloys, Ref 10 and 21 are highly recommended.

Basic Requirements for Hardening by Heat Treatment

As stated above, practically all nonferrous alloys respond readily to annealing treatments, but only a relatively small portion of the total number respond significantly to hardening by heat treatment. With the exception of titanium, common high-use aluminum, copper, and magnesium alloys are not allotropic; thus, they do not respond in the same way as steels when subjected to heating and cooling treatments (see Chapter 2).

At one time, technologists thought that ancient civilizations had developed a method for hardening copper alloys that had become a lost art. Copper articles removed from tombs showed a relatively high degree of hardness. However, it was proved that the impurities in these copper articles were responsible for their hardness. This hardness had evidently developed over a period of centuries by the precipitation mechanism rather than intentionally induced by the people who made them.

Solubility and Insolubility. In many instances, two or more metals, alloyed above their melting temperature, are completely soluble in each other, which continues through the solid solution range. For example, copper and tin alloy readily to form an entire "family" of bronzes. For these alloys, the solid solutions that form at elevated temperature remain completely stable at room temperature or below. Such an alloy, therefore, can be hardened only by cold working.

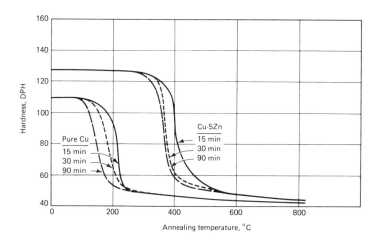

Fig. 2 Hardness as a function of annealing temperature for pure copper and a Cu-5Zn alloy, using three different annealing times. Both materials were originally cold rolled at 75 °F (25 °C) to a 60% reduction in thickness. Source: *Heat Treatment, Structure and Properties of Nonferrous Alloys*, p 29, 1982.

On the other hand, many alloys contain phases or constituents that are readily soluble at elevated temperature, but are far less soluble or insoluble at room temperature, which is the basic requirement for hardening by heat treatment.

General Heat Treating Procedures

Alloys with the characteristics described above are hardenable at least to some degree. The metallurgical principles responsible for this phenomenon are the same (or at least closely related) for all of the nonferrous alloys. Likewise, methods of processing are basically the same. However, temperature and time cycles cover a wide range, depending not only on alloy composition but also on whether the alloy is in the wrought or cast condition.

As a rule, the highest possible increase in mechanical properties for a given alloy is accomplished in two distinct heating operations. First, the alloy is heated to a temperature just below (say at least 50 °F or 28 °C) its solidus (beginning of melting) temperature. Holding time at temperature depends on the solubility of the various phases and whether the alloy is wrought or cast. It must be considered that the primary purpose of this operation is to affect a solid homogeneous solution of all phases. Consequently, this part of the operation is commonly known as solution treating or homogenizing.

Because the structures of castings are relatively heterogeneous, a much longer time (at least twice as long) is required at the solution treating temperature compared to a wrought product of similar composition.

Cooling From the Solutionizing Temperature. After solid solution has been attained, the alloy is cooled as quickly as possible to room temperature (most often by water quenching) to arrest the structure obtained by heating. The required cooling rate is more critical for some alloys than for others. In addition, section thickness markedly affects cooling rate; still air is sometimes adequate.

At this stage, the alloy is in the solution treated condition and, in most cases, is fully annealed (usually as soft as it can be made). However, the alloy is in an unstable condition because the phase or phases that possess a low degree of solubility at room temperature are virtually all dissolved.

Aging (Precipitation Hardening). As soon as room temperature is reached, these phases start to precipitate out of solid solution in the form of fine particles within the crystals and at the grain boundaries. At one time, these particles were considered as submicroscopic, but the use of the electron microscope has changed this.

In most alloys, precipitation occurs very slowly at room temperature and is even more sluggish at lower temperatures. In fact, this action in most alloys can be completely arrested by lowering to subzero temperatures. However, if the temperature of the solution treated alloy is raised, the action is greatly increased and can be completed in a relatively short time.

Aging Cycles. The precipitation or dispersion of fine particles tends to key the crystals together at their slip planes so that they are more resistant to deformation; thus, the alloy is stronger, harder, and less ductile. For each alloy, there is an optimum time and temperature. For the majority of nonferous alloys, this is from about 325 to 650 °F (165 to 345 °C), although some alloys require higher aging temperatures.

As is true for most heat treating operations, precipitation is a function of time and temperature. For example, a specific alloy might assume its maximum strength by aging for 5 h at 325 °F (165 °C). Overaging in either time or temperature is highly detrimental and decreases strength of the alloy.

Specific Alloy Systems

When precipitation hardening mechanisms are discussed, one is usually referring to nonferrous metals, although there are some iron-based alloys that are hardened by the precipitation mechanisms (see Chapter 7). Compositions and typical uses of some well-known alloys of aluminum, copper, magnesium, and nickel that respond readily to precipitation hardening are given in Table 1.

The above discussion has, thus far, been quite general. We shall now take two commercially important alloys—one from the aluminum-copper system and one from the copper-beryllium system—and demonstrate more clearly the mechanism involved in solution treating and aging. As previously stated, time and temperature cycles vary greatly for different alloys, but the principles are basically the same for all precipitation hardening.

To accomplish this, it is necessary to use phase diagrams. For more information on the use of phase diagrams, see Chapters 2 and 9.

Aluminum-Copper Alloys

The first example that demonstrates the conditions which must exist for precipitation hardening is the partial aluminum-copper phase diagram presented in Fig. 3.

In this phase diagram, percent copper is plotted on the horizontal axis, and temperature is plotted on the vertical axis. The left side of the diagram (up to about 4.5% Cu) is the significant part of the diagram for showing the mechanisms of solution and aging.

First, note that there is terminal solid solution (denoted as Al) on the left of the diagram (shaded area). The curved line ABC shows the maximum solid solubility of copper in aluminum for the temperature range of 0 to 1200 °F (32 to 650 °C). The horizontal line at 1018 °F (550 °C) is the eutectic temperature. If the copper content is greater than that shown by line AB, the solid aluminum is saturated with copper and a new phase, $CuAl_2$, appears.

Table 1 Precipitation hardenable alloys
Source: Metals Engineering Institute Course 1, Lesson 12, American Society for Metals, p 2, 1977.

Alloy type	Principal alloying elements, %	Typical uses
Aluminum-based alloys		
2014 (formerly 14S)	4.4 Cu, 0.8 Si, 0.8 Mn, 0.4 Mg	Forged aircraft fittings, aluminum structures
2024 (formerly 24S)	4.5 Cu, 1.5 Mg, 0.6 Mn	High-strength forgings, rivets, and structures
6061 (formerly 61S)	1 Mg, 0.6 Si, 0.25 Cu, 0.25 Cr	Furniture, vacuum cleaners, canoes, and corrosion-resistant applications (used in ASM's geodesic dome)
7075 (formerly 75S)	5.5 Zn, 2.5 Mg, 1.5 Cu, 0.3 Cr	Strong aluminum alloy; used in aircraft structures
Copper-based alloys		
Beryllium bronze (beryllium copper)	1.9 Be, 0.2 Co or Ni	Surgical instruments, electrical contacts, non-sparking tools, springs, nuts, gears, and other heavy duty applications
Aluminum bronze	10 Al, 1 Fe	
Magnesium-based alloys		
AM100A	10 Al, 0.1 Mn	Tough, leakproof sand castings
AZ80A	8.5 Al, 0.5 Zn, 0.15 Mn min	Extruded products and press forgings
Nickel-based alloys		
Rene 41	0.1 C, 19 Cr, 10 Co, 10 Mo, 3 Ti, 1.5 Al, 1.5 Fe, 0.005 B	High temperature applications where strength is very important (up to about 1800 °F)
Inconel 700	0.13 C, 15 Cr, 30 Co, 3 Mo, 2.2 Ti, 3.2 Al, 1 Fe	
Udimet 500	0.1 C, 19 Cr, 19 Co, 4 Mo, 3 Ti, 2.9 Al, 4 Fe (max.), B trace	

Therefore, when aluminum containing about 4.5% Cu is heated to around 1000 °F (540 °C), it becomes a single-phase stable alloy at that temperature. Because of the similarity in size of the copper and aluminum atoms, the alloying is hardening by substitution (trading atoms from their lattice position) rather than by interstitial alloying, as explained in Chapter 2 for the alloying of iron and carbon. Thus, at various points in the aluminum lattice, copper atoms are substituted for aluminum atoms. When the alloy is quenched in water (or otherwise quickly cooled), there is insufficient time for the copper to precipitate as $CuAl_2$. According to Fig. 3, copper at room temperature is nearly insoluble in aluminum (shown as less than 1.0%). At this point, the alloy is in an unstable condition.

Aging. Under these very unstable conditions, some precipitation of $CuAl_2$ occurs at room temperature. This can be checked by first measuring the hardness in the as-quenched condition (essentially as soft as the alloy can be made), then continuing to take hardness measurements at regular intervals. In a matter of hours, the alloy will register an increase in hardness caused by precipitation of $CuAl_2$ particles. The rate of hardness at room temperature, however, gradually diminishes, and after a time (say a few days), action practically ceases. However, if the as-quenched alloy is heated to about

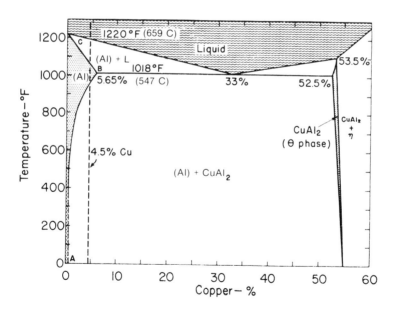

Fig. 3 Part of the aluminum-copper phase diagram. The kappa phase, bounded by ABC, is a solid solution of copper in aluminum; $CuAl_2$ precipitates from this phase on slow cooling or on aging after solution treatment. Source: Metals Engineering Institute Course 1, Lesson 12, American Society for Metals, p 3, 1977.

340 °F (170 °C) for several hours, the copper atoms regain enough mobility to effect the optimum amount of precipitation. Therefore, maximum attainable strength for the specific alloy is developed quickly by this artificial aging (purposely heating above room temperature).

Overaging. There is an optimum time and temperature for aging of any specific alloy. When overaging takes place—in terms of either time or temperature, or both—there is a rapid drop in hardness. This is clearly shown in Fig. 4. It is also obvious from Fig. 4 that a fair-sized area exists that shows "maximum hardness' (zone 2), which indicates that there is some variation in the time-temperature cycles.

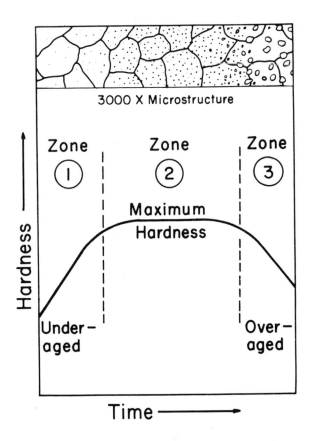

Fig. 4 Schematic aging curve and microstructure. At a given aging temperature, the hardness of aluminum-copper alloys increases to a maximum, then drops off. Source: Metals Engineering Institute Course 1, Lesson 12, American Society for Metals, p 4, 1977.

Changes in microstructure also are included in Fig. 4. Fine particles are precipitated in zone 2, whereas coarse particles (over precipitation) are shown within the grains in zone 3, where a definite decrease in hardness exists.

Copper-Beryllium Alloys

Copper-beryllium alloys hold a unique position among all other engineering alloys that are in use today. Copper by itself is used extensively throughout the electrical industry because it conducts electric current better than any other metal (excluding silver). Its resistance to the flow of current is low. By itself, however, copper does not have a great deal of strength. It is too soft to use in most structural applications. To impart some strength to copper, other elements are alloyed with it. One of the more potent of these is beryllium. The alloys of copper containing beryllium are among the strongest known of the copper alloy family. This high strength combined with good electrical conductivity makes copper-beryllium alloys (also known as beryllium bronze or beryllium copper) particularly suited for electrical applications, as in contacts and relays which must open and shut a great number of times. Often these switches must open and close many times each minute, and during years of continuous service, they may be stressed a billion times or more. Beryllium bronze is highly suited for applications of this type. The same part may also serve as a structural member in the actual relay to hold it together and to support other parts. Heat treated copper-beryllium alloys also have been used extensively in aircraft and aerospace applications.

Phase Diagrams. A partial copper-beryllium phase diagram is presented in Fig. 5. This diagram can be interpreted using the guidelines described in Chapter 2, and in much the same manner as the aluminum-copper system in Fig. 3. Figure 5 only shows beryllium contents up to 10%. The remainder of the phase diagram is not included because there are no commercial copper-based alloys in use with beryllium content above 2½%. Alloys with about 2% Be are used where high strength is required. A ½% beryllium alloy is used where high electrical conductivity is the main objective. However, to get good aging response, about 2% Ni or Co must be added to such an alloy.

This phase diagram for the copper-beryllium system is slightly more complex than those presented previously, and for that reason we will not discuss it here in any detail. The important feature to note is the all-important decreasing solubility of beryllium in the alpha (α) phase with decreasing temperature (large shaded area on the left of Fig. 5).

The precipitation process in the copper-beryllium alloy system is complicated by the beta phase changing into gamma. However, for the alloy under discussion the reader is interested only in the left portion of the phase diagram. Note that the primary requirement for precipitation hardening has been met; that is, beryllium has a very low solubility in copper at low temperatures and a relatively high solubility at elevated temperatures.

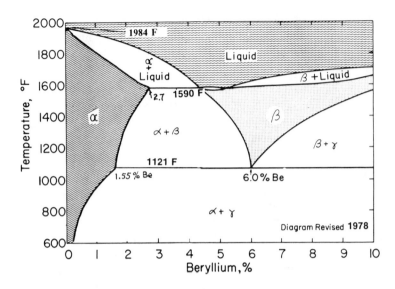

Fig. 5 Beryllium-copper phase diagram. The alpha phase holds about 1.55% Be at 1121 °F (605 °C) and about 2.7% at 1590 °F (865 °C). Decreasing solubility of the beryllium causes precipitation of a hard beryllium-copper phase on slow cooling or on aging after solution treatment. Source: Metals Engineering Institute Course 1, Lesson 12, American Society for Metals, p 10, 1977.

Solution Treating and Aging. In considering heat treatment, the shape of the hardness versus aging time curves for alloys with varying amounts of beryllium should be considered. Figure 6 shows the aging curves for several copper-beryllium alloys containing from 0.77 to 4.0% Be. These alloys were all solution treated at 1470 °F (800 °C) and aged at 660 °F (350 °C) for various lengths of time, as shown in Fig. 6.

Note that an alloy containing as little as 0.77% Be does not harden as a result of aging. However, when the beryllium content is raised to 1.32%, a significant hardness increase takes place after 16 h. By increasing the beryllium content to 1.82%, a great deal of hardening takes place after 4 h, and the hardness stays at essentially the same level for the duration of the aging period. A beryllium content of 2.39% raises the peak hardness attained for this alloy. By increasing the beryllium content to 4%, still higher maximum hardnesses can be attained (Fig. 6).

Several interesting observations can be made from examining Fig. 6. First, note that at zero aging time (the solution treated condition), alloys with higher beryllium content are harder. This is known as solid solution hardening. The atoms of the parent lattice are displaced in making an adjustment. As a result of the shifting of position, the hardness of the alloy

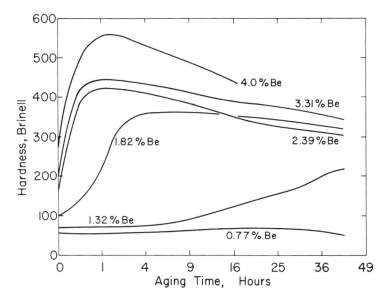

Fig. 6 Precipitation-hardening curves of beryllium-copper binary alloys. As the percentage of beryllium increases, the aging time required to reach maximum hardness is shortened, and the maximum hardness is increased. These alloys were quenched from 1470 °F (800 °C) and aged at 660 °F (350 °C) for the times shown. Source: Metals Engineering Institute Course 1, Lesson 12, American Society for Metals, p 11, 1977.

increases. A similar but less intensely stressed condition results in solid solution hardening. When beryllium atoms are present in the copper lattice, the copper atoms are forced to move slightly out of their normal position to accommodate the somewhat different size of the beryllium atom. This accommodation or slight shifting in position gives rise to a moderate increase in hardness. When more beryllium atoms are present, then more copper atoms have to shift. The result is a higher hardness as the number of beryllium atoms increases.

Attaining Maximum Hardness. As the beryllium content is increased up to 2.39%, the time required to reach the maximum hardness decreases. Thus, aging for 36 h does not allow time enought to produce maximum hardness in a 1.32% beryllium alloy. With 1.82% beryllium, however, the maximum hardness occurs after approximately 5 h. With 2.39% Be or more, the maximum hardness occurs after only 1 h. It can also be seen from this diagram that by increasing the beryllium content above 2.39%, the rate of aging is not significantly increased. The 3.31 and 4% beryllium alloys reach a maximum hardness at about the same time as the 2.39% alloy. By looking at the phase diagram of the beryllium-copper alloy system, the reason for this can be seen. The maximum solid solubility of beryllium in copper is just about 2.7% (at a temperature of 1590 °F, or 865 °C). Thus, the 3.31% alloy contains more than the maximum amount of beryllium that can be dissolved in copper. This means that the increased hardness of the 3.31 and 4% alloys is not due to any additional aging effects but is simply caused by the presence of undissolved beta phase, as may be predicted from the phase diagram. This increase in hardness on account of undissolved precipitate is known as dispersion hardening. The undissolved precipitate is harder than the matrix surrounding it, and the final hardness of the mixture depends on how much undissolved precipitate is present, as well as its manner of distribution.

References

1. *Glossary of Metallurgical Terms and Engineering Tables,* American Society for Metals, 1979

2. *Elements of Metallurgy,* Course 1, Lesson 11, Metals Engineering Institute, American Society for Metals, 1977

3. Zapffe, C.A., *Stainless Steels,* American Society for Metals, 1949

4. *Heat Treatment of Steel,* Course 10, Lesson 3, Metals Engineering Institute, American Society for Metals, 1981

5. *Heat Treatment of Steel,* Course 10, Lesson 6, Metals Engineering Institute, American Society for Metals, 1977

6. *Metals Handbook,* H.E. Boyer, editor, Vol 11, 8th edition, American Society for Metals, 1976

7. *Heat Treater's Guide: Standard Practices and Procedures for Steel,* P.M. Unterweiser, H.E. Boyer, and J.J. Kabbs, editors, American Society for Metals, 1982

8. *Heat Treatment of Steel,* Course 10, Lesson 7, Metals Engineering Institute, American Society for Metals, 1977

9. *Energy Conservation in Heat Treating and Forging Operations,* Course 20, Lesson 4, Metals Engineering Institute, American Society for Metals, 1980

10. *Metals Handbook,* Vol 4, 9th edition, American Society for Metals, 1981

11. ASM Seminar on Energy Management in Metalworking, paper by H.E. Boyer, published by British Columbia Chapter, American Society for Metals, 1981

12. *Metals Handbook,* Taylor Lyman, editor, Vol 2, 8th edition, American Society for Metals, 1964

13. *Heat Processing Technology,* Course 6, Lesson 15, Metals Engineering Institute, American Society for Metals, 1978

14. *Stainless Steels,* Course 18, Lesson 2, Metals Engineering Institute, American Society for Metals, 1978

15. *Principles of Heat Treating,* Course 41, Lesson 14, Metals Engineering Institute, American Society for Metals, 1969

16. *Tool Steels,* AISI Steel Products Manual, American Iron and Steel Institute, Washington, DC, 1970

17. *Applications and Properties of Ferrous Castings,* Course 17, Lesson 8, Metals Engineering Institute, American Society for Metals, 1981

18. *Gray and Ductile Iron Castings Handbook,* Gray and Ductile Iron Founder's Society, Cleveland, 1971

19. *Carburizing and Carbonitriding,* American Society for Metals, 1977

20. *Heat Treatment of Steel,* Course 10, Lesson 9, Metals Engineering Institute, American Society for Metals, 1979

21. Brooks, Charlie R., *Heat Treatment, Structure and Properties of Nonferrous Alloys,* American Society for Metals, 1982

22. *Elements of Metallurgy,* Course 1, Lesson 12, Metals Engineering Institute, American Society for Metals, 1977

Index